Higher Geography

How to Pass

SECOND EDITION



HIGHER

# Geography

Dr Bill Dick

Consultant editor: Sheena Williamson

HODDER GIBSON
AN HACHETTE UK COMPANY

**Dedication:** In memory of my friend and fellow geographer Leslie Mill (BD).

The Publishers would like to thank the following for permission to reproduce copyright material.

**Photo credits** © lazyllama – Fotolia (p. 74)

**Acknowledgements**

Every effort has been made to trace all copyright holders, but if any have been inadvertently overlooked, the Publishers will be pleased to make the necessary arrangements at the first opportunity.

Although every effort has been made to ensure that website addresses are correct at time of going to press, Hodder Gibson cannot be held responsible for the content of any website mentioned in this book. It is sometimes possible to find a relocated web page by typing in the address of the home page for a website in the URL window of your browser.

Hachette UK's policy is to use papers that are natural, renewable and recyclable products and made from wood grown in well-managed forests and other controlled sources. The logging and manufacturing processes are expected to conform to the environmental regulations of the country of origin.

Orders: please contact Bookpoint Ltd, 130 Park Drive, Milton Park, Abingdon, Oxon OX14 4SE. Telephone: (44) 01235 827827. Fax: (44) 01235 400454. Email education@bookpoint.co.uk Lines are open from 9 a.m. to 5 p.m., Monday to Saturday, with a 24-hour message answering service. Visit our website at www.hoddereducation.co.uk. Hodder Gibson can also be contacted directly at hoddergibson@hodder.co.uk



This second edition published in 2019 by
Hodder Gibson, an imprint of Hodder Education
An Hachette UK Company
211 St Vincent Street
Glasgow, G2 5QY

Impression number 5 4 3 2 1

Year 2023 2022 2021 2020 2019



Cover photo © aleciccotelli – stock.adobe.com
Illustrations by Aptara
Typeset in 13/15 Cronos Pro Light by Aptara, Inc.
Printed in Spain
A catalogue record for this title is available from the British Library.
ISBN: 978 1 5104 5241 1

# Contents

# Introduction

## The aims of this guide

This *How to Pass Higher Geography* guide will not cover the topics in as great a detail as your class texts and support notes. However, when revising it will give a useful guide to the main points that examiners look for in answers. You do not have to remember all of the points in each chapter but you should try to gain a basic knowledge of the topics covered.

The guide contains a series of key points and key words, and throughout there is a variety of questions and example answers provided from previous examinations. These answers have been marked, and comments on the quality of the answers and the marks obtained are provided. There are references in the text and in sample answers to case studies. These are intended as a guide. Additional detail on case studies should be provided in course notes by your class teacher.

When preparing for the exam you will need to cover physical and human environments and any two global issues from a choice of four. The global issues will be chosen for you by your teacher.

Your teacher may be able to help you predict possible questions or topics in the external examination that you will sit. However, this kind of prediction is not a perfect science and should be used with caution.

This guide will concentrate on the external examination component of the overall assessment. It will not cover or advise on detail of the assignment component of the assessment since that will be dealt with internally within your school or college.

There is another point that you should note: You will find that as you go through this text there will be an amount of overlap between chapters, especially between the chapters on atmosphere, global climate change, hydrosphere and river basin management. Indeed, in the advice provided by the Scottish Qualifications Authority (SQA), there is guidance on how the study of different topics can be combined. Your teacher will refer you to published SQA documents that will indicate how this can be achieved in your course studies. The overlap of topics has been deliberately included in this guide to reinforce SQA guidelines.

The aim of this text is not to replace core study. Some background information has been included to provide context to enhance the quality of your answers in the exam, and this has been signposted throughout the text by a blue-tinted panel.

## Course assessment structure

The following information is taken from the Higher Geography course specification provided by SQA.

The course assessment has three components.

| Component | Marks | Scaled mark | Duration |
|---|---|---|---|
| 1 Question paper 1 – Physical and human environments | 100 | 50 | 1 hour and 50 minutes |
| 2 Question paper 2 – Global issues and geographical skills | 60 | 30 | 1 hour and 10 minutes |
| 3 Assignment | 30 | not applicable | 1 hour and 30 minutes |

Question paper 1 represents 46 per cent, Question paper 2 represents 27 per cent and the Assignment represents 27 per cent, giving a total of 100 per cent.

## Question paper 1: Physical and human environments

This question paper has 100 marks out of a total of 190 marks. This is scaled by SQA to represent 46 per cent of the overall marks for the course assessment.

This question paper enables you to demonstrate your skills, knowledge and understanding of the physical and human environments sections of the course.

In this question paper you have an opportunity to:

- use a wide range of geographical skills and techniques
- describe, explain, evaluate and analyse complex geographical issues, using knowledge and understanding. This should be both factual and theoretical, and cover physical and human processes and interactions within geographical contexts on a local, regional and global scale.

Personalisation and choice is possible through case studies and areas chosen for study.

This question paper has two sections:

- **Section 1: Physical environments**

  **Key topics** include: atmosphere; hydrosphere; lithosphere; and biosphere.
- **Section 2: Human environments**

  **Key topics** include: population; rural land degradation and management; and urban change and management.

Each of these sections is worth 50 marks and consists of extended-response questions. You will have to answer all the questions in each section.

You will have **1 hour and 50 minutes** to complete this question paper.

## Question paper 2: Global issues and geographical skills

This question paper has 60 marks out of a total of 190 marks. This is scaled by SQA to represent 27 per cent of the overall marks for the course assessment.

This question paper enables you to demonstrate your skills, knowledge and understanding from across the global issues and geographical skills sections of the course.

In this question paper you will have an opportunity to:

- use a wide range of geographical skills and techniques
- describe, explain, evaluate and analyse complex geographical issues, using knowledge and understanding – both factual and theoretical – of the physical and human processes and interactions within geographical contexts on a local, regional and global scale.

Personalisation and choice is possible through case studies and areas chosen for study.

This question paper has two sections:

- **Section 1: Global issues** is worth 40 marks and consists of extended-response questions. You have to choose two out of four questions. Each question is worth 20 marks.
  **Key topics include:** river basin management; development and health; global climate change; and energy.
- **Section 2: Application of geographical skills** is worth 20 marks and consists of a mandatory extended-response question.
  In answering this question, you are being asked to apply the geographical skills you have acquired during the course. The skills assessed in the question include mapping skills and the use of numerical/graphical information.
- You have **1 hour and 10 minutes** to complete this question paper.

## Component 3: The assignment

Component 3 is an internal assignment conducted within **1 hour and 30 minutes**, with the use of specified resources, under a high level of supervision and control for which evidence will be submitted to SQA for external marking. It will be marked out of 30 and form 27 per cent of the total mark.

The assignment will have a greater emphasis than the question paper (component 1) on assessing your skills.

It will allow you to demonstrate your skills and your knowledge and understanding of a geographical topic or issue of your choice.

You will:

- identify a geographical topic or issue
- carry out research, which should include fieldwork where appropriate
- be asked to show how suitable are the methods and/or how reliable are the sources used
- process data using a range of information gathered
- draw on detailed knowledge and understanding of the topic/issue
- analyse information from a range of sources
- reach a conclusion supported by a range of evidence
- demonstrate your skills of communicating information

## Skills, knowledge and understanding for the course assessment

| Geographical skills |
|---|
| The following skills are assessed in contexts drawn from across the course:<br>Mapping skills:<br>• interpretation and analysis<br>• using maps, including Ordnance Survey maps, in association with photographs, field sketches, cross sections/transects<br>Research skills including fieldwork skills:<br>• gathering<br>• processing<br>• interpreting<br>• evaluating<br>Using numerical and graphical information which may be presented in the following ways:<br>• statistical<br>• graphical<br>• tabular |

| Physical environments |
|---|
| In relation to physical environments, candidates:<br>• develop and apply geographical skills and knowledge and understanding<br>• develop and apply knowledge and understanding of the processes at work and interactions with human environments on a local, regional and global scale<br>Content sampled in this section of the question paper: |
| **Atmosphere** |
| • global heat budget<br>• redistribution of energy by atmospheric and oceanic circulation<br>• cause, characteristics and impact of the Inter-Tropical Convergence Zone |
| **Hydrosphere** |
| • formation of erosional and depositional features in river landscapes:<br>  ◦ V-shaped valley<br>  ◦ waterfall<br>  ◦ meander<br>  ◦ oxbow lake<br>• hydrological cycle within a drainage basin<br>• interpretation of hydrographs |
| **Lithosphere** |
| • formation of erosional and depositional features in glaciated landscapes:<br>  ◦ corrie<br>  ◦ arête<br>  ◦ pyramidal peak<br>  ◦ U-shaped valley<br>  ◦ hanging valley<br>  ◦ ribbon lake<br>  ◦ drumlin<br>  ◦ esker<br>  ◦ terminal moraine |

- formation of erosional and depositional features in coastal landscapes:
  - wave-cut platform
  - headland and bay
  - cave
  - arch
  - stack
  - spit
  - bar
  - tombolo

**Biosphere**

- properties and formation processes of podzol, brown earth and gley soils

**Human environments**

In relation to human environments, candidates:

- develop and apply geographical skills and knowledge and understanding
- develop and apply knowledge and understanding of the processes and interactions at work within urban and rural environments in developed and developing countries
- evaluate the impact/effectiveness of management strategies

Content sampled in this section of the question paper:

**Population**

- methods and problems of data collection
- consequences of population structure
- causes and impacts of forced and voluntary migration

**Rural**

- impact and management of rural land degradation related to a rainforest or semi-arid area
- rural land-use conflicts and their management related to either a glaciated or coastal landscape

**Urban**

- the need for management of recent urban change (housing and transport) in a developed- and in a developing-world city
- management strategies employed
- impact of management strategies

**Global issues**

In relation to global issues, candidates:

- develop and apply geographical skills and knowledge and understanding
- develop and apply knowledge and understanding of significant global geographical issues that demonstrate the interaction of physical and human factors and evaluate strategies adopted in the management of these issues

Candidates study two of the four global issues. An appreciation of sustainable development should permeate the global issues studied.

Content sampled in this section of the question paper:

**River basin management**

- physical characteristics of a selected river basin
- need for water management
- selection and development of sites
- consequences of water control projects

| Development and health |
| --- |
| • validity of development indicators<br>• differences in levels of development between developing countries<br>• a water-related disease: causes, impact, management<br>• primary health-care strategies |
| **Global climate change** |
| • physical and human causes<br>• local and global effects<br>• management strategies and their limitations |
| **Energy** |
| • global distribution of energy resources<br>• reasons for changes in demand for energy in both developed and developing countries<br>• effectiveness of renewable and non-renewable approaches to meeting energy demands and their suitability within different countries |

# Examination advice

- When revising, you can use this guide together with past examination specimen papers to revise topic by topic, for example Hydrosphere.
- The question paper will be set and marked by SQA, and conducted under conditions specified by them.

## Some revision tips

- For the actual examination, prepare your notes in sections.
- Try to work out a schedule for a study programme that includes the sections of the syllabus you intend to study.
- Organise your notes into checklists and revision cards.
- Try to avoid leaving your studying to a day or two before the exam.
- Also try to avoid cramming your studies on the night before the examination, especially staying up late to study.
- Practise drawing diagrams that may be included in your answers, for example corries or pyramidal peaks.
- Make sure you know the examination timetable and note the dates and times of your examinations.
- Give yourself plenty of time by arriving early for the examination, well equipped with pens, pencils, erasers and so on.

## Some tips for the exam

- If you are asked for a named country or city, make sure you include details of any case study you have covered.
- Avoid vague answers when asked for detail, for example avoid vague terms such as 'dry soils' or 'fertile soils' and instead try to provide more detailed information in your answer, such as 'deep and well-drained soils' or 'soils rich in nutrients'.
- If you are given data in the form of maps, diagrams and tables in the question, make sure you refer to this information in your answer to support any points of view that you give.
- The length of your answer should be guided by the number of marks available for that question.

- Watch your time and do not spend too much on any particular answer, thus leaving yourself short of time to finish the paper. Make sure that you leave yourself sufficient time to answer all of the questions.
- If you have any time left in the exam, use it to go back over your answers to see if you can add anything by way of additional text, including more examples or diagrams that you may have omitted.
- Make sure that you read the instructions to the questions carefully and that you avoid needless errors such as answering the wrong sections, failing to explain when asked to do so or perhaps omitting to refer to a named area or case study.
- One technique that can be helpful, especially when answering long questions, is to 'brain storm' for possible points for your answer. You can write these down in a list at the start, then as you write your answer you can double-check your list to ensure that you put in as much as you can. This avoids coming out of the exam and being annoyed that you forgot to mention an important point.

# Common exam errors

## Lack of sufficient detail

- Higher case study answers, especially in questions with high marks, often lack sufficient detail.
- Many candidates fail to provide sufficient detail in their answers by omitting reference to specific examples or omitting to elaborate or develop points they have made in their answer.
- Remember that you have to give more information in your answers to gain a mark than in the old exam.

## Irrelevant answers

- You must read the question instructions carefully so as to avoid giving answers that are irrelevant to the question. For example, if you are asked to explain and you simply describe, you will not score marks; if you are asked for a named example and you do not provide one, you will forfeit marks.

## Statement reversals

- Occasionally questions involve opposites, for example some answers would say 'death rates are high in developing countries due to poor health care' and then go on to say 'death rates are low in developed countries due to good health care'. Avoid doing this. You are simply stating the reverse of the first statement and are not making a separate point. A better second statement might be that 'high standards of hygiene, health and education in developed countries have helped to bring about low death rates'.

## Repetition

- You should be careful not to repeat points you have already made in your answer. These will not gain any further marks.
- You may feel that you have written a long answer, but it may contain the same basic information repeated again and again. Unfortunately, these statements will be ignored by the marker.

## Listing

- If you give a simple list of points rather than fuller statements in your answer you may lose marks. For example, in a 5-mark question you will obtain only 1 mark for a list.

## Bullet points

- The same rule applies to a simple list of bullet points. However, if you give bullet points with some detailed explanation you could achieve full marks.

# Section 1 Physical environments

## Chapter 1.1
## Atmosphere

## Global heat budget

The amount of heat or energy received by the Earth from the Sun varies throughout different parts of the Earth, due mainly to the effect of **latitude**.

> *Key point* !
>
> You should be able to explain the distribution of the amount of solar energy that is absorbed by the Earth. This is known as the Earth's heat budget.

### The Earth's heat budget

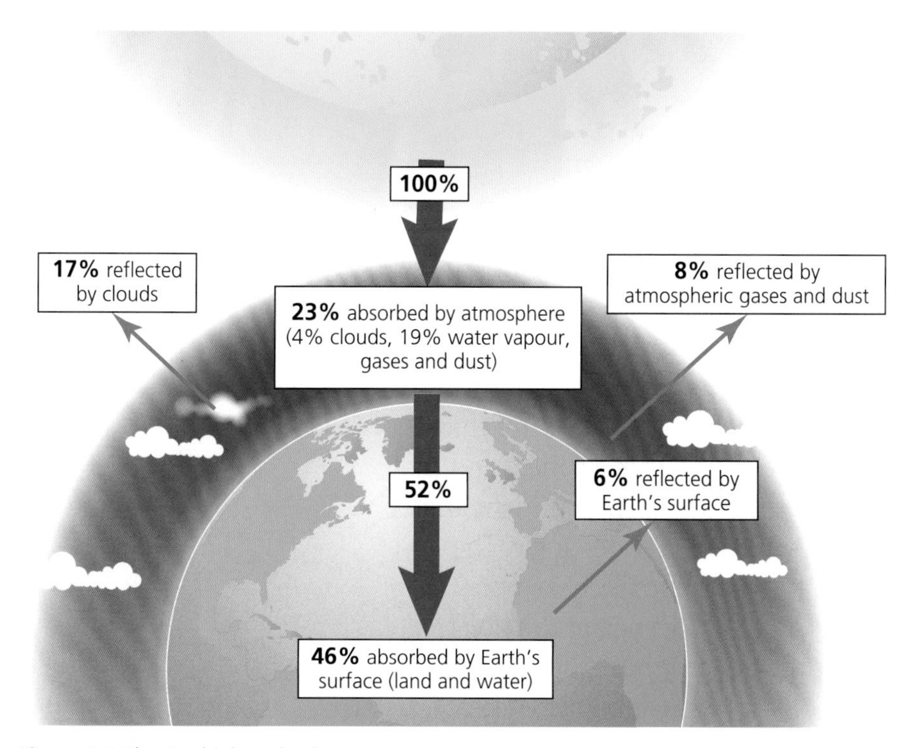


**Figure 1.1** The Earth's heat budget

- Figure 1.1 shows a summary of the energy that the Earth receives from the Sun and how it is distributed.
- The Earth is heated by energy in the form of solar rays from the Sun. Some of this energy is absorbed directly by land and water, some is interrupted by clouds and dust in the atmosphere and some is reflected back into space before even reaching the Earth's surface.
- For every 100 units of energy that the Earth receives, 31 units are reflected back into space: from clouds (17), gases and dust (or the atmosphere) (8) and from the Earth's surface (6).

- This reflected energy is called the Earth's **albedo**.
- A further 21 units are absorbed by clouds (4), water vapour, dust and various gases (19).
- The remaining 46 units are absorbed by the Earth's surface (land and water).

## Factors that affect the amount of sunlight reflected from the Earth's surface

The most important points to note include:

- Energy absorbed by the Earth causes the Earth's temperature to rise. Energy radiates back from the surface into the atmosphere, where it is absorbed by clouds and gases. These clouds and gases are heated and they in turn radiate energy, some of which is returned to the Earth's surface.
- The return of this energy from the atmosphere once again heats the surface and helps to maintain surface temperatures.
- The system of energy movement can be interrupted by the emission of gases such as carbon dioxide from factories, for example. Energy that would be reflected back into space is trapped and this causes the atmosphere to heat up further, creating what is called the **greenhouse effect**.
- Most of the energy that heats the atmosphere actually comes from the Earth's surface. This happens through conduction, that is, energy rising from the Earth's surface and through latent heat given out when evaporated water rises from the surface and condenses in the atmosphere.

*Key point* 

You should be able to describe the factors that affect the amount of sunlight reflected from the Earth's surface.

Other factors include the variation in the Earth's **insolation** due to the effect of latitude on the distribution of **solar energy**:

- Places at the equator or between the tropics are always hotter than places at higher latitudes. This is because the Earth is a sphere and the Sun's rays strike the areas around the centre of the Earth at right angles.
- At higher latitudes the rays strike the surface at a wider angle. The net effect of this is that surfaces nearer the equator receive more insolation – that is, the Sun's heat – than surfaces nearer the poles.
- The insolation striking the surface at the equator heats up a smaller surface area than the same amount of insolation at higher latitudes.

## The Earth's movement around the Sun

There is a net gain of solar energy towards the tropics and a net loss of energy towards the poles because:

- The Earth revolves around the Sun. Therefore the Sun is overhead at the equator and each of the tropics at different times of the year.
- When the Sun apparently appears directly overhead at the equator, this is called the **maxima**; it is also termed the **equinoxes**.

*Key point* !

You should be able to explain why there is a net gain of solar energy towards the tropics and a net loss of energy towards the poles.

- When the Sun is directly over the tropics, this is termed the **solstices**. During the winter solstice in the northern hemisphere, the amount of insolation received at the North Pole is zero.
- At the poles there are alternately six months of light and six months of darkness.

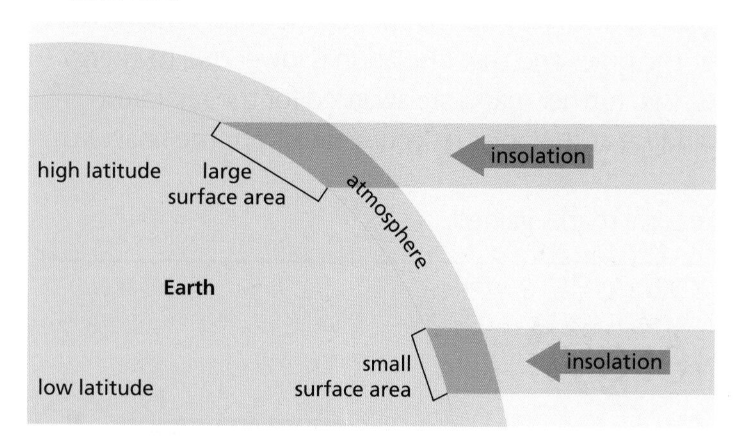


**Figure 1.2** Variation in the Earth's insolation

*Remember that reflection rates vary: ice, snow and water have high rates while forest cover has a low rate.*

## Quick test

Why are lower latitudes warmer than higher latitudes? In effect, why are equatorial areas very warm while polar areas are very cold? This is due to lower latitudes receiving more of the Sun's energy than polar regions. You can use a diagram if asked to explain this.

## Example

With the aid of an annotated diagram or diagrams, explain why there is a net gain of solar energy in tropical latitudes and a net loss towards the poles. **10 marks**

### Sample answer

*Between the Equator and the Tropics the sun's rays have less atmosphere to travel through (✓) so less energy is lost through atmospheric absorption and reflection. (✓) The rays of the sun cover a smaller area so are more concentrated (✓) so the intensity of insolation is higher here. (✓)*

*At the Poles the earth is tilting away from the sun so the sun's rays have to travel further through the atmosphere (✓) so energy covers a larger area so insolation is lower here. (✓)*

*At the tropics areas of dense vegetation like the Rainforest absorb radiation (✓) whereas at the Poles areas covered in snow and ice reflect the incoming radiation back into the atmosphere. (✓)*

*The rays have to spread out and travel through more atmosphere so are more diluted as the earth tilts away from the sun. Rays hit the surface at a narrow angle so travel through less atmosphere and are more concentrated.*

➪

## Comment and marks

The first sentence, referring to less energy being lost through atmospheric absorption and reflection, merits two marks – one for description and the second for explanation. The cover of a smaller area at the equator plus reference to the intensity of insolation deserves two marks – one for description and the second for explanation. The tilt of the Earth at the poles and that insolation is lower due to energy covering a larger area is worth another two marks. Two further marks are awarded for the rainforest absorbing radiation being contrasted with snow and ice at the poles reflecting radiation. The final two sentences are repetition so no marks awarded.

A total of **8 marks out of 10** available are the total marks gained.

# Global transfer of energy

You must be able to describe the role of atmospheric circulation in the redistribution of energy over the globe. This is called **global transfer of energy**.

Global transfer of energy is due to the following:

- In addition to the vertical transfer of energy between the atmosphere and the Earth's surface, energy is also transferred between the equator and the poles.
- Areas north and south of latitude 38° receive less solar energy than areas between latitudes 38° N and 38° S.
- In the higher latitudes, more energy is emitted from the surface than is absorbed. Nearer the tropics more energy is absorbed by the surface than is emitted from it.
- Consequently there is a deficit in solar energy north and south of latitudes 38° and a surplus of solar energy in areas between 38° N and 38° S.
- If this situation remained static, areas near the tropics would become hotter while those further north and south would become colder. That this does not happen is due to the transfer of energy from areas of surplus to areas of deficit. This transfer of energy is known as **atmospheric circulation**.

*Key point* 

You should be able to describe and explain the pattern of atmospheric circulation and global winds and their contribution to the global transfer of energy using an appropriate diagram. Figure 1.3 shows this three-cell model of atmospheric circulation.

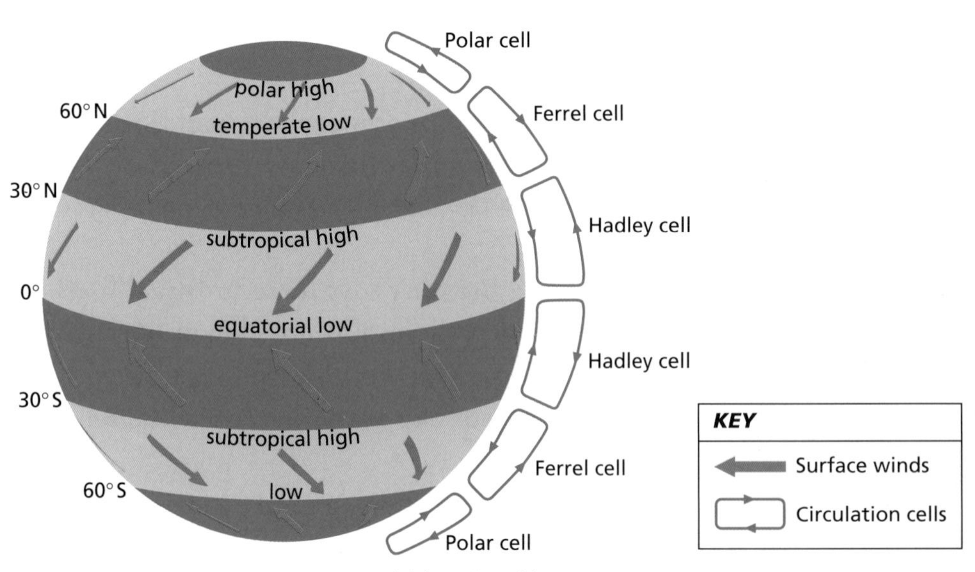


**Figure 1.3** Atmospheric circulation model (Ferrel's cells)

## Atmospheric circulation

Atmospheric circulation happens because:

- At the equator, the energy at the surface of the Earth heats the air immediately above it. This air expands, becomes less dense and rises to a higher altitude, creating a zone of low pressure. At the higher altitude the temperature is colder and therefore the air cools, becomes more dense and begins to fall. Differences in pressure between the surface and upper atmosphere create a wind.
- Due to the rotation of the Earth and the giving out of latent heat, the cooler, more dense air flows both northwards and southwards, and as it becomes even cooler and denser, it falls as an area of high pressure. This falling air creates a high pressure zone around 30° N and 30° S. This circulation of air forms cells both north and south of the equator called **Hadley cells**.
- Some of this air moves from the high pressure area towards lower latitudes due to the movement of the Earth. In these lower latitudes, the air is heated and begins to rise into higher altitudes where it is cooled, creating a zone of low pressure. Thus a circulation pattern of air occurs at the poles, similar to that above the tropics, called **polar cells**.
- A third cell termed a **Ferrel cell** forms due to the temperature differences between the first two cells at the tropics and at the poles. Warm air from the Hadley cell at the tropics feeds into the higher latitudes, while colder air from the polar cells feeds into the lower latitudes.
- This leads to the transfer of energy from the warmer lower latitudes to the higher colder altitudes, and transfer of colder air from the colder higher latitudes to the warmer lower latitudes.

**Quick test ?**

Referring to Hadley and polar cells, explain how the Ferrel cell is formed.

## Pressure and wind patterns

A simple pattern of distribution of **pressure belts** is shown in Figure 1.4.

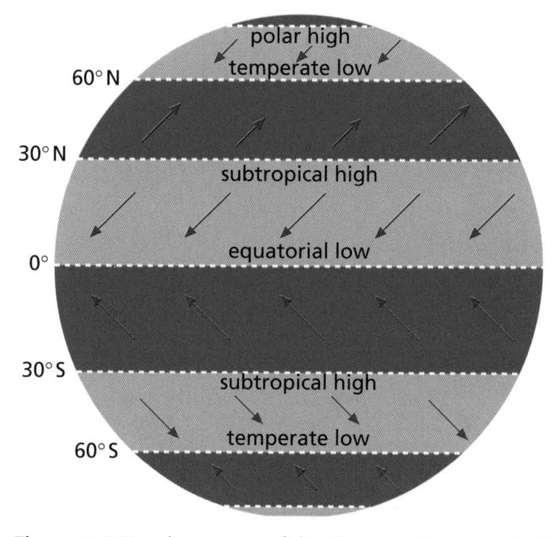


**Figure 1.4** Simple pattern of distribution of pressure belts

This pressure belt pattern changes throughout the year due to several factors, including:

- As air is heated it expands and rises, leaving the pressure near the surface low since there is less air to create pressure. As the air rises it cools and becomes denser and falls.
- In areas where the air is cooler it is denser and this creates an area of high pressure.
- Areas north and south of the equator are relatively cooler and are areas of high pressure.
- The air at the poles is also colder and denser and this forms areas of high pressure.
- Air blows from high to low pressure, creating winds. Air will blow outwards from the high subtropical areas towards the equator and towards the relatively lower pressure areas between the subtropics and the poles.

Reasons for the patterns include:

- The position of the Sun changes during the seasons due to the Earth revolving around the Sun. This affects the position of the pressure belts, which changes. During summer in the northern hemisphere the Sun is overhead at the Tropic of Cancer. The high pressure belt moves further north and the other pressure belts also move northwards.
- In December, when the Sun moves overhead at the Tropic of Capricorn, the pressure belts move further southwards. These movements affect world patterns of temperature and winds throughout the year.
- The rotation of the Earth also affects pressure belts. This movement tends to deflect air from the poles towards the equator. Air moving from high to low pressure is deflected to the right in the northern hemisphere and to the left in the southern hemisphere.
- Land and sea masses affect pressure and wind patterns. The rate of heating and cooling varies greatly over land and sea areas. This creates distortions in the pattern of pressure belts, with a corresponding effect on winds.
- The flow of wind throughout the Earth is affected by landscape features such as mountains, which deflect winds from their path.
- There are large belts of fast-moving winds traversing the globe at high altitudes of between 10,000 and 12,000 metres. The pattern is wave-like due to the influence of temperature and pressure differences between land and sea areas. The wave patterns have been termed **Rossby waves**.
- There are streams of very fast-moving air known as **jet streams** within these waves. These occur due to differences in temperature between the polar, subtropical and equatorial **air masses**. These waves and jet streams contribute greatly to the movement of energy throughout the world.

The features of the contribution of the main wind patterns throughout the Earth to the global transfer of energy include:

- There are two main wind belts in each hemisphere; namely, the **trade winds** and the **westerlies**.
- Trade winds are found between latitudes 30° N and 30° S and are caused by the movement of air from the high pressure subtropical zones towards the low pressure zone at the equator.
- The westerly winds flow pole-wards out from the high subtropical pressure areas towards the middle latitude areas in the northern and southern hemispheres. The trade winds are more constant and predictable in this part of the world.
- There are smaller belts of winds flowing outwards from the poles in an easterly direction in both hemispheres, which occur only in winter in the northern hemisphere. In the southern hemisphere, the pattern is not confined to any particular season.
- Areas further inland are less influenced by these winds.

*Key point* 

You should be able to explain the contribution of the main wind patterns throughout the Earth to the global transfer of energy.

## The pattern of the world's ocean currents

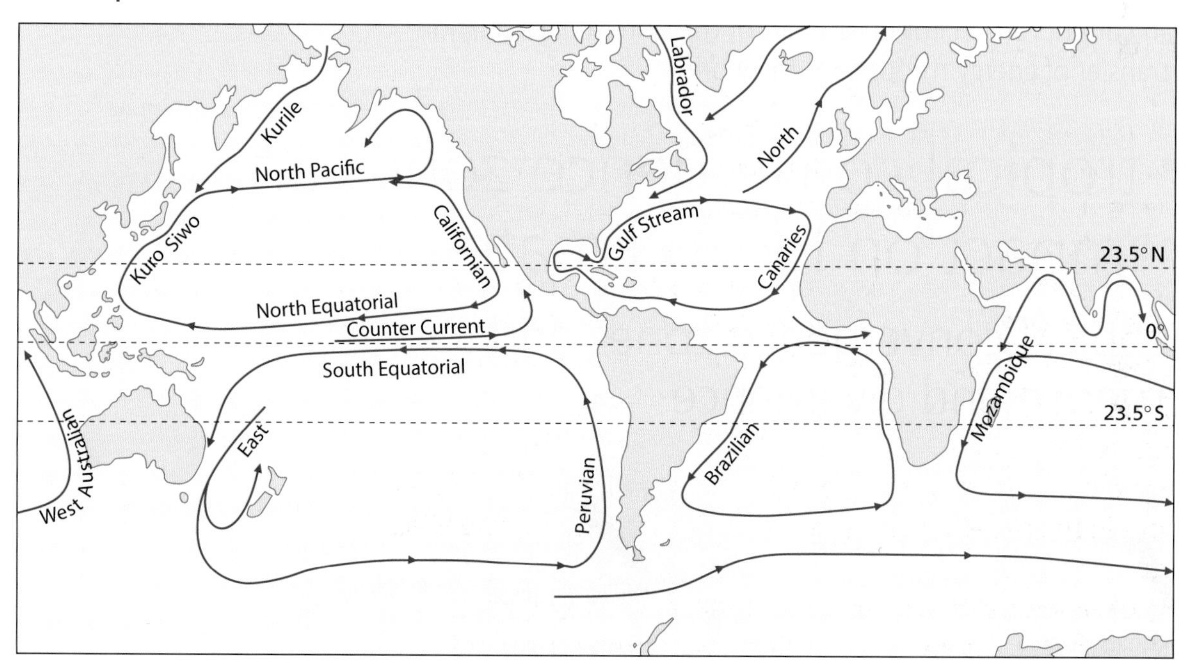


**Figure 1.5** Pattern of world's ocean currents

- Some 71 per cent of the Earth's surface is covered by water and 29 per cent of the surface is land. This has an important influence on the transfer of energy since water is a much more efficient store of heat than land.
- The oceans warm more slowly than land but are heated to a greater depth. Heat is redistributed due to the flow of ocean currents.
- Because the waters nearer the equator receive more heat than those near the poles, warm water flows outwards from the equatorial regions towards higher latitudes.
- In turn, colder water from the poles flows towards warmer regions, creating a circulatory system. The flow is disrupted and distorted by the effect of the Earth's rotation and the distribution pattern of the world's land masses, creating the pattern of ocean currents that exist at present.

*Key point* 

You should be able to explain how ocean currents affect the circulation of global energy. Figure 1.5 shows this pattern of the world's ocean currents.

- The reasons for this pattern include:
  - The pattern of the world's ocean currents is closely linked to the distribution of the world's main pressure belts and wind patterns.
  - The land masses disrupt the flow of the currents – otherwise the pattern would be fairly straightforward and in both hemispheres there would be three groups of currents: equatorial, subtropical and sub-arctic/Antarctic.
  - The presence of continental land masses distorts the flow of the currents, producing the pattern shown in Figure 1.5.
  - Winds blowing over these currents would assist in the flow of warm water to cooler areas and cooler water to warmer areas.
  - Due to the Earth's rotation, winds in the northern hemisphere are deflected to the right and in the southern hemisphere to the left, helping to create the pattern of currents shown in Figure 1.5.
  - The nature of the current – whether it is warm or cold – and the type of wind – whether it is onshore or offshore – has a vital effect on climatic conditions on the land masses.
  - The pattern of ocean currents has a great influence on the temperature patterns throughout the world through the different seasons.
  - Ocean currents are an important part of the system of circulation and transfer of energy throughout the world.

# Inter-tropical convergence zones and their impact on local climate

## Inter-tropical convergence zones and convergence and divergence

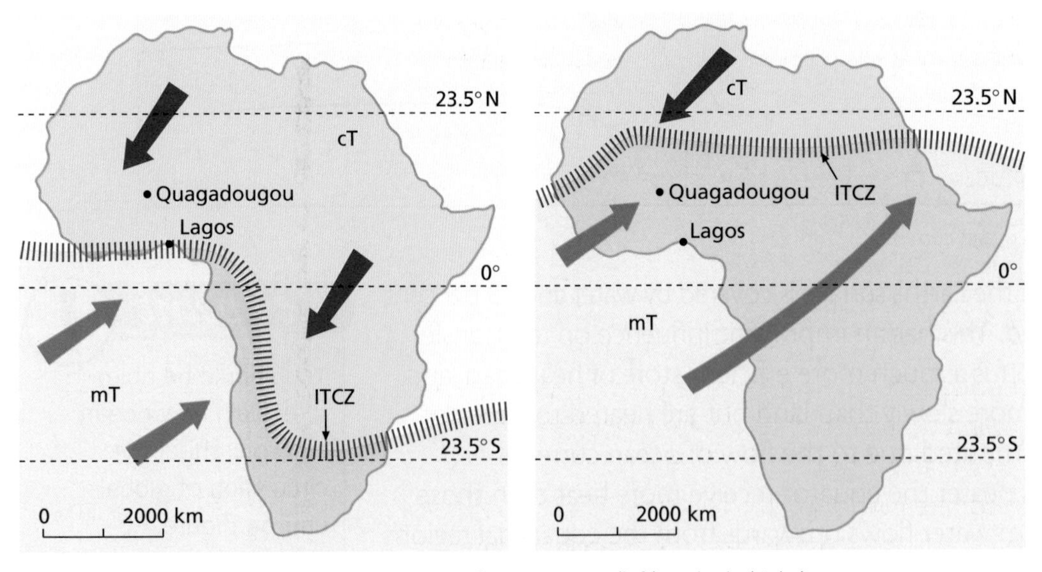


**Figure 1.6** Selected air masses and fronts over Africa – January (left) and July (right)

- A **zone of convergence** is where winds meet, and **zone divergence** is an area where the winds go in different directions. Converging winds include the trade winds, which meet at the equator in a zone termed the **Inter-Tropical Convergence Zone (ITCZ)**.
- Winds that flow southwards from the polar areas converge with air flowing pole-wards from the subtropical high belts. Convergence of easterly winds with the westerlies occurs along the polar front.
- Two zones of divergence occur in the subtropical high pressure zones where the winds are usually fairly light.
- These zones are referred to as the **horse latitudes**. Climatic conditions of the horse latitudes consist generally of clear skies, abundant sunshine, low rainfall and calm or very light variable winds.
- In these zones, and in the ITCZ, the movement of air is vertical or convectional. In the ITCZ the air tends to rise along the inter-tropical front.
- The zone of convergence moves northwards and southwards with changes in the Sun's angle of declination during the seasons. This shift in the ITCZ affects the **climate** of areas in these latitudes, especially **rainfall patterns**.
- When air masses from different source regions meet, the air at the edge begins a process in which the colder air forces warm air upwards and condensation takes place in the upper parts. The place where this occurs is called the **front**.
- Where the trade wind belt of the northern latitudes meets the trade wind belt of the southern latitudes within the equatorial belt, an inter-tropical front is formed. The weather associated with this front depends on whether the front has formed over the oceans or the continents:
  - Air masses converging towards the inter-tropical front over oceans are moist in the lower layers and relatively dry at higher levels. At convergence there is some instability and large cumulus clouds appear, which eventually leads to intense shower and thunder conditions.
  - Within equatorial areas, the vertical movement of air through convection produces convectional rain. This belt of equatorial rainfall moves north or south with the annual changes in the Sun's declination of about 5° latitude either side of the equator.

Key point 

You should know about inter-tropical convergence zones and convergence and divergence, as shown in Figure 1.6.

## The ITCZ and West Africa

How and why are there variations in rainfall in West Africa? What impact does the ITCZ have on climatic conditions in West Africa? These variations are affected by the following:

- In the northern hemisphere from March to July the ITCZ moves northwards across West Africa, bringing heavy convectional rainfall. This rainy season lasts only about two months.
- At the northern edge conditions are much drier, where the air meets dry air from the interior of the continent.

- As the ITCZ moves further southwards in winter, the drier continental tropical air is drawn southwards, giving drier conditions in the north-western areas.
- In some years there has been an apparent shift in the movement of the ITCZ, which has resulted in less rainfall than normal in the areas north of the equator. This, combined with a southward extension of the subtropical high pressure from the Sahara, has resulted in long periods of **drought**.
- The climate of equatorial Africa is best described as high rainfall throughout the year with high temperatures and a small annual **range of temperature**.
- The main physical impact that the ITCZ has is its effect on the climate of West Africa.
- As the ITCZ shifts its position throughout the year, it brings wet and dry seasons to the area.

## Key words and associated terms

**Air mass:** A large volume of air that is composed of the same properties.
**Albedo:** The amount of reflectability of surfaces on the Earth such as land, ice and water.
**Climate:** The average conditions of weather usually taken over a period of 35 years.
**Drought:** A long period without rainfall. Notice that droughts do not necessarily just happen in desert areas.
**Front:** The boundary between two air masses. If the air on one mass is warmer than the air being replaced, the front is termed a 'warm front'. If the air is colder than the air being replaced, the front is termed a 'cold front'.
**Insolation:** The amount of heat taken in from the Sun.
**Inter-Tropical Convergence Zone (ITCZ):** A broad zonal trough of low pressure in equatorial latitudes where air converges at the equator.
**Latitude:** The distance between the equator and the poles, measured in degrees.
**Pressure belts:** Patterns of atmospheric circulation systems of either high or low atmospheric pressure.
**Rainfall pattern:** The distribution of rainfall throughout the year.
**Range of temperature:** The difference between the highest temperature and the lowest temperature in a year.
**Solar energy:** Any form of energy originating from the Sun.

Chapter 1.2
# Hydrosphere

Hydrology is the study of the water within the Earth, whether it is in the atmosphere, on the surface or underground. The movement of that water, the impact of it on the land and how the movement may be interrupted are also important aspects of this particular topic.

## Erosional and depositional features in river landscapes

### V-shaped valley

- V-shaped valleys are mainly found in the upper course of the river. In this area the water flows naturally downhill and erodes downwards (vertical erosion).
- Boulders, stones and rock particles are scraped along the channel bed, creating steep valley sides using the processes of hydraulic action, corrasion and corrosion.
- **Hydraulic action** is when the power of the water forces air into gaps in the banks and weakens them so they eventually collapse.
- **Corrasion** is the wearing away of the riverbed and banks by the load hitting against them.
- **Corrosion** is when the water dissolves minerals from the rocks and washes them away.
- Over time, as the river cuts down, steep sides are attacked by weathering called freeze–thaw. This breaks up and loosens the soil and rock.
- Gradually the loosened materials move down the valley sides (slope transport) creating the V-shape. The river then transports the materials away and the river channel becomes wider and deeper creating a V-shaped valley.

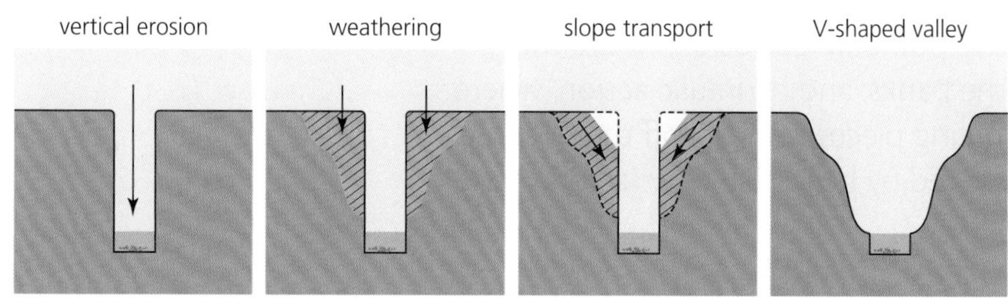


**Figure 1.7** The formation of V-shaped valleys

# Waterfalls

- Waterfalls are found where there is a layer of hard rock on top of a softer rock.
- The softer rock is eroded more quickly. This is called **differential erosion**.
- The water is powerful and erodes the softer rock by hydraulic action. This is the force of the water hitting the rock. The softer rock is worn away and the hard rock is undercut, leaving an overhang of hard rock. At the base, a plunge pool is formed.
- The overhanging rock is left unsupported and falls into the plunge pool. Rock fragments swirling around deepen the plunge pool by **abrasion.** Hydraulic action also deepens the plunge pool.
- This process is repeated over a long period of time and the waterfall retreats upstream forming a steep-sided gorge.

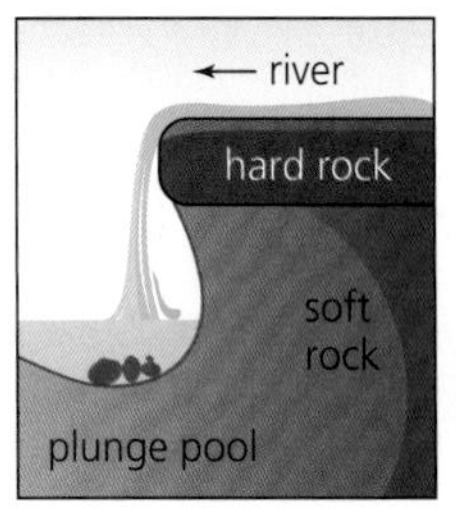

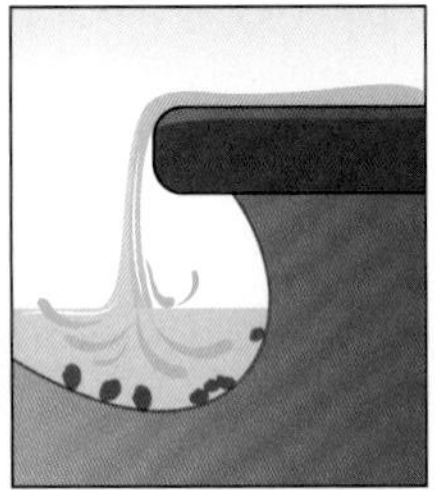

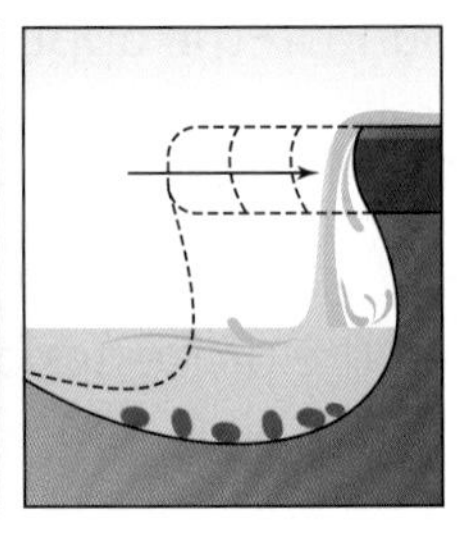

**Figure 1.8** The formation of waterfalls

# Meanders

- Meanders are common in the middle and lower courses of a river. They are formed as a result of the processes of erosion and deposition.
- Due to lateral erosion, the river widens and spreads across the valley floor. Pools develop where the flow is deepest, and riffles develop in the shallow flow.
- The fastest current is found on the outside of the bend because the depth of the water there is deeper, so there is less friction resulting in higher speeds. The outside bend of the river is eroded to form a river cliff.
- The processes happening at this point are corrosion, where the load of the river wears away the banks, and hydraulic action, where water gets into small cracks forcing pieces to break off the riverbed and banks. The material is removed by helicoidal flow (a corkscrew movement).

- The slowest current is on the inside of the bend as it has more contact with the riverbed. This causes the river to deposit some of the load it carries and forms a river beach or slip-off slope.
- Continuous erosion of the outer bank and deposition on the inner bank form a meander in the river.

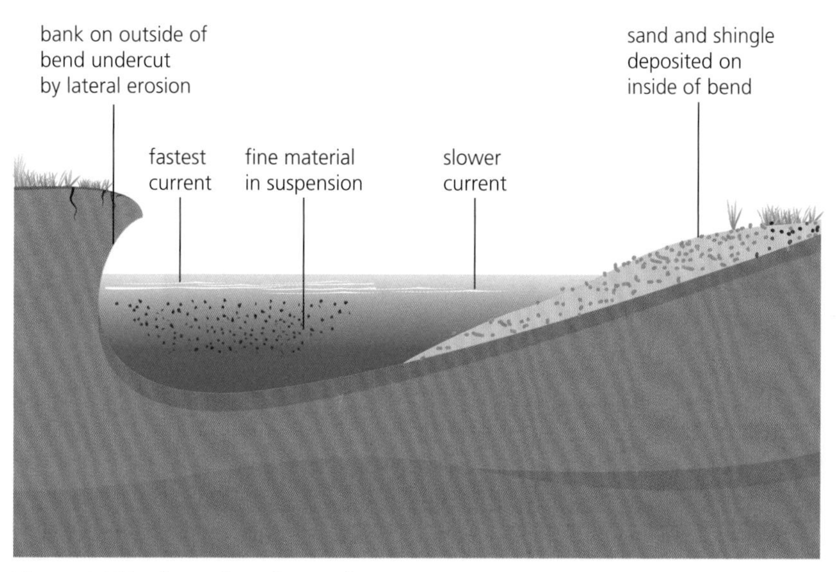


**Figure 1.9** The formation of meanders

# Oxbow lakes

- Erosion of the outside of the meander means that the neck of land becomes narrower and narrower over time.
- During periods of flood or high discharge, the river breaks through the neck, creating a straighter, easier channel for the river to flow through. This creates an oxbow lake.
- Deposition over time causes the lake to be sealed off from the main channel of the river.
- Over time the water can dry out but since this is the river's **floodplain**, at times of high rainfall, water might once again fill up the oxbow lake.

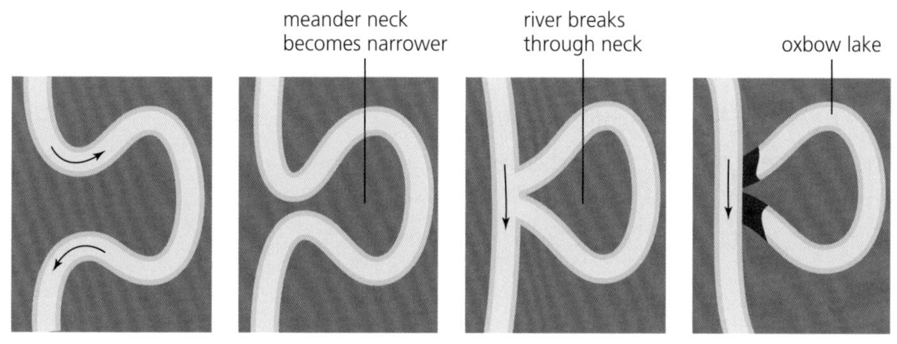


**Figure 1.10** The formation of an oxbow lake

# The hydrological cycle

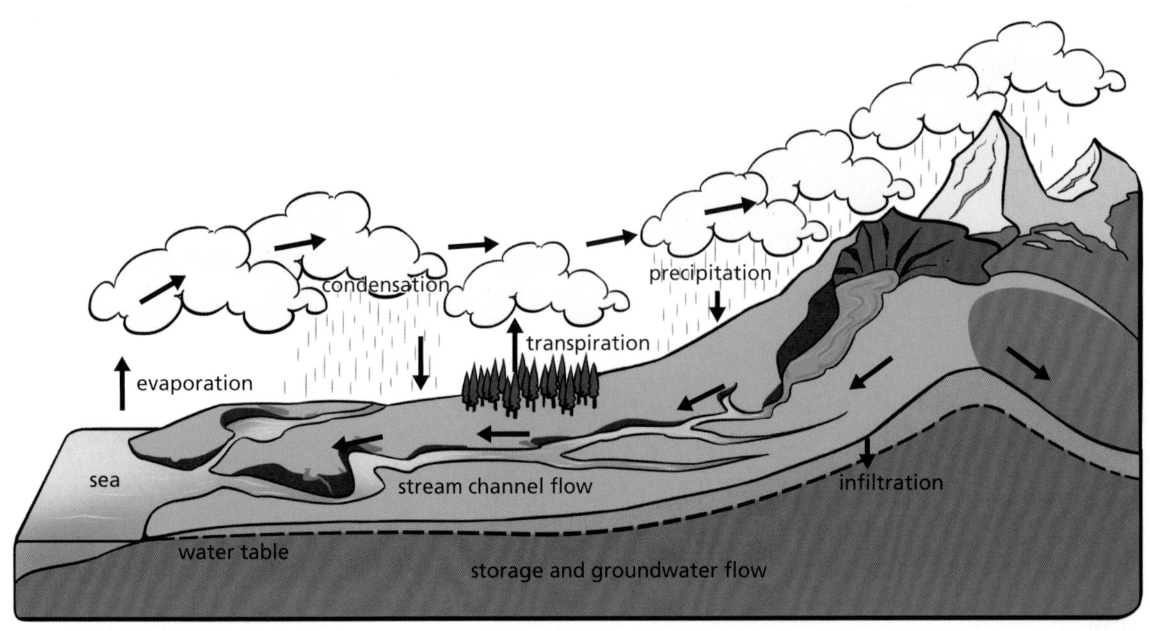


**Figure 1.11** Hydrological cycle

The intricate process of the movement of water back and forth between land, oceans and the atmosphere is called the **hydrological cycle**.

1 Water exists on the surface in the form of oceans, seas, lakes, rivers and streams. It also exists in the atmosphere as rain and water vapour and underground as seepage within rock structures and in underground streams and lakes. The surface water can pass into the atmosphere through evaporation/**evapotranspiration** and can be carried by winds and eventually returns to the surface as rain or snow.
2 Water also exists on the surface as ice or snow, for example at the poles or at high altitudes. Water on the land may be returned to oceans and seas through rivers or streams.
3 The hydrological cycle works as a closed system, in that there is a finite amount of water in the atmosphere, lithosphere and **hydrosphere** (land and water areas). This amount remains constant.
4 The system is powered by energy from the Sun. Within the system the amount of water in the various components can and does vary, especially when the system is interrupted.

*Key point* 

You should be able to draw a diagram to show the global hydrological cycle, naming and describing the main elements within the diagram.

# How balance is maintained within the hydrological cycle

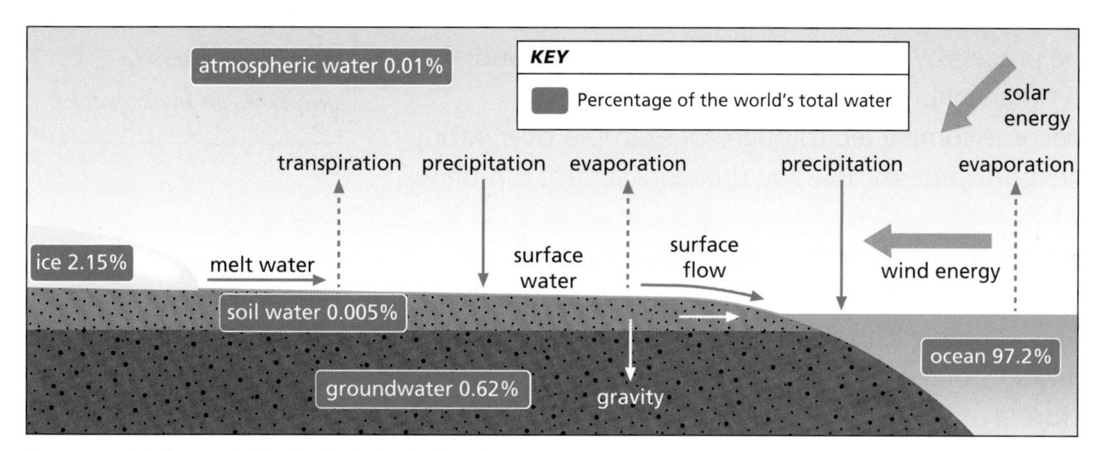


**Figure 1.12** Balance within the hydrological cycle

There is a continuous movement of water between the different parts of the system through the processes of evaporation, transpiration and **precipitation**.

## Evaporation and precipitation

- Evaporation and precipitation rates vary widely across the world. The energy that creates evaporation is obtained from the Sun.
- Areas where temperatures are high have high evaporation rates, for example the equatorial and tropical areas, especially over the oceans.
- Warm winds help water to evaporate more than cold calm conditions and move air from place to place.
- Areas that have dry conditions, such as deserts, have relatively low evaporation rates since there is very little surface water to evaporate.

## River drainage basins

- **Soil water** and **groundwater storage** varies according to changes in precipitation, evaporation, transpiration, **infiltration** and local geology.
- **Drainage basins** are the total areas, known as **catchment areas**, which drain into rivers.
- A drainage basin system includes **stored water**, which is water held within the system in lakes or in the soil. **Water transfer** refers to water percolating or infiltrating rock strata.
- The return of that water may be interrupted by some being absorbed by soils and by vegetation such as trees.
- Each basin is separated from neighbouring basins by a **watershed**. This feature is the area of high ground dividing basins where the surface water flows in different directions.

> **Hints & tips**
>
> *When discussing balance, remember the following equation:*
>
> *Balance is maintained when Inputs = Outputs*
>
> ***Inputs:***
> - *rainfall*
> - *rivers and streams flowing into the seas*
> - *groundwater seepage into the soil*
> - *evaporation from the sea and rivers.*
>
> ***Outputs:***
> - *transpiration from plants*
> - *evaporation from land surfaces.*
>
> *The balance is affected if there are changes to the basic system through, for example, interruption in the run-off through water storage systems such as dams, or changes in world temperature patterns through the greenhouse effect.*

- During periods of heavy rainfall or low temperatures the soil and subsoil may become too saturated or impenetrable, resulting in an increase in surface run-off. Similarly if rainfall is low, certain flow channels may become dry if the soil is permeable.
- **Infiltration** is the process by which water soaks into the ground.
- If rainfall is greater than infiltration, surface run-off occurs.
- Soils that have become compacted through, for example, overgrazing may also become more impermeable and therefore surface run-off will increase.
- Infiltration is faster in sandy soils than in clay soils.
- **Interception** is the storage of rainwater above the ground surface, mostly in vegetation. Some of this water may never make it into soil water or groundwater storage because of **transpiration** from leaves or direct **evaporation**.
- Cultivation of forests in some areas will have a significant impact on the hydrological cycle in a river basin.
- **Groundwater** may lie either on the surface or within certain rocks for some considerable time. If the rock is porous or permeable the water may be held within the rock or slowly seep through.

Hints & tips

*Write a list of the highlighted words and use them in answers to questions on hydrographs.*

# Hydrographs

## Techniques and features

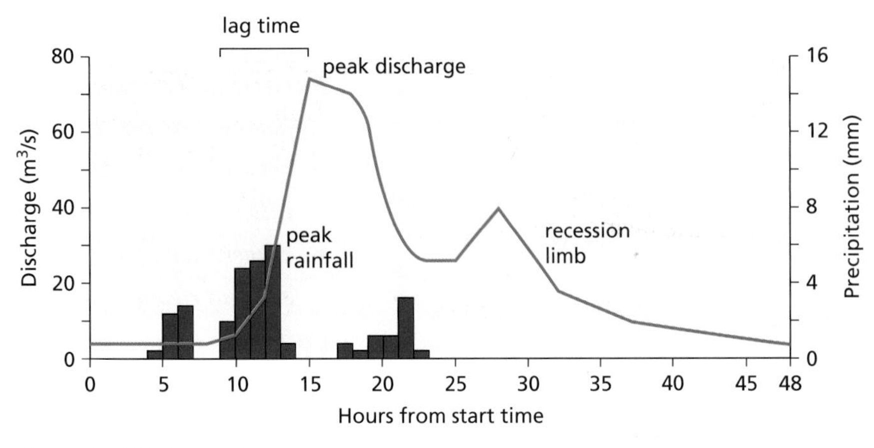


**Figure 1.13** Hydrograph

- You should know the key features of a hydrograph.
- You should also be able to describe and explain patterns shown on a river hydrograph.

The rate of flow or discharge of water within a river basin can be measured or recorded by a graphic technique known as a hydrograph, as shown in Figure 1.13.

Important features of a hydrograph that you should be able to refer to include:

- **Total rainfall:** The amount of rainfall that has fallen over a specific length of time, usually several days.
- **Time:** The time over which the run-off or discharge is measured and recorded.
- **Discharge:** The amount of water that has been discharged by the basin within the specified time scale, measured in cumecs ($m^3$/second).

*Write a list of the highlighted words to the left and on page 17 and try to use them in answers to questions on hydrographs.*

- **Rising limb:** Indicates how quickly waters begin to rise.
- **Peak flow:** The maximum discharge during a storm period.
- **Time lag:** The difference in time between the height of a storm and the maximum flow of the river.
- **Recessional** or **falling limb:** Indicates the speed at which the water level in the river declines after its peak flow.
- **Base flow:** The normal level of a river.
- **Quick flow:** The water that is fed into the river due to overland run-off.

## Storm hydrographs

A storm hydrograph is one that displays two basic features:

- **Feature 1:** The rainfall from a rain storm, which is shown by a bar graph.
- **Feature 2:** The river discharge before, during and after the rain storm, which is shown by line graphs.

In effect, the storm graph indicates how a river responds to a rain storm.

## Analysing hydrographs

When analysing a hydrograph, note the following:

- Discharge during a storm does not increase immediately since only a small amount of the rain will fall directly into the river channel.
- Water will reach the river from the overland flow. This is the surface run-off and it will subsequently be supplemented by water from through-flow. This is shown on the graphs in Figure 1.14 on the next page. You should note this in any analysis you carry out.
- The rising limb shows the overland flow and the falling limb, which is less steep, indicates that there is still water in the system from through-flow, surface water and water in stream channels. Note these too in analysis.
- Note the **lag time** and discharge. Rivers that have a short lag time and a high discharge are more likely to flood than rivers with a long lag time and a low discharge.
- Depending on the characteristics of river basins in terms of size, shape, relief, geology and so on, hydrographs for two basins receiving the same amount of rainfall can be very different. The graphs in Figure 1.14 illustrate this.

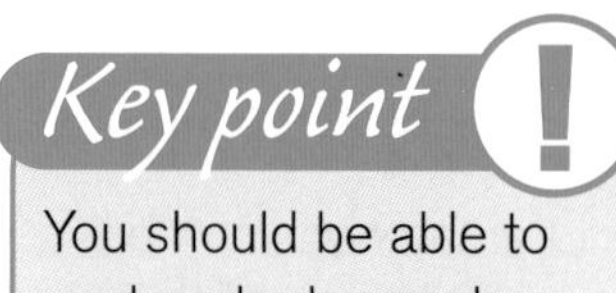

**Key point**

You should be able to analyse hydrographs and also compare hydrographs for two different river basins.

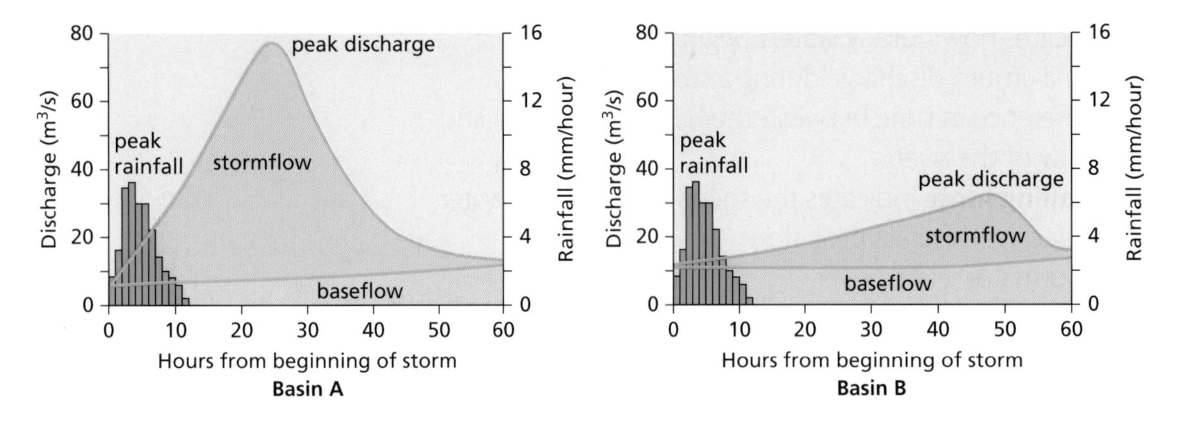


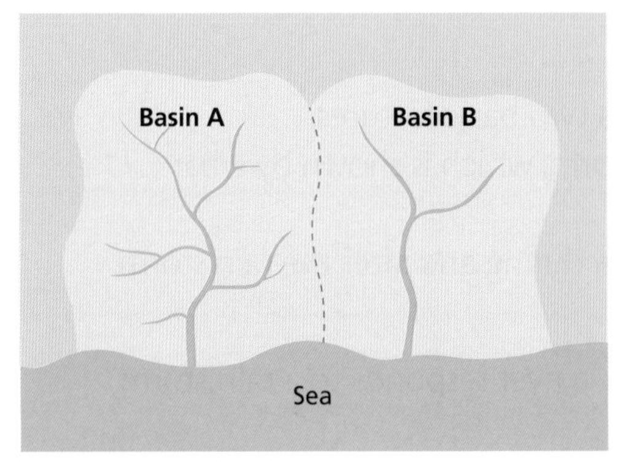


**Figure 1.14** Two selected flood hydrographs/river basins

## Analysis of Figure 1.14 hydrographs

- Drainage basin A has a much higher density than that of drainage basin B. Comparison of the hydrographs for each basin shows that for basin A the lag time is much shorter (30 hours) than for basin B (55 hours).
- The peak discharge for A is much higher (140 cumecs) than for basin B (50 cumecs).
- The rising limb for B is much less steep than for A, as is the recession (falling) limb.
- Apart from the difference in density of drainage within the two basins, it is quite possible that there is some considerable variation in the relief pattern, where the pattern may be much steeper in basin A than in B.
- The soil and rock types may also vary, thereby affecting the overland flow patterns and perhaps the through-flow patterns.
- Further analysis might reveal that there are some differences between the vegetation present within the two basins, for example there may be more woodland present in B than in A, which will affect discharge.
- The net result of these differences in hydrographs and basin characteristics is that the river in basin A is more likely to flood than the river within basin B.
- Note, however, that without more detail on the characteristics of both systems, flood predictions are rather speculative.

# The effects of changes to a drainage system on different hydrographs

- Hydrographs are affected by the area of the basin since the higher the basin size the greater the discharge. The amount of discharge is also related to the amount of rainfall. Therefore the peak flow will be higher in larger basins.
- If the slopes are steep, the infiltration will be less and therefore the peak flow will be greater. If the basin has flatter slopes, there will be more infiltration and this will result in lower peaks.
- The lag time is also affected by steepness of slope.
- Quick flow will increase where the rainfall is high and the infiltration is low. The discharge will also be affected by local vegetation, the number of tributary streams and whether the drainage basin has been affected by human land use.
- Town and city hydrographs are different from those of rural environments since there are much more impermeable surface areas, such as pavements and roads, surface drains and gutters, in urban areas.
- There may also be less vegetation to interrupt the run-off.

## Key point !

In addition to being able to describe and explain hydrographs, you should be able to explain the effects that changes to a drainage system, such as increase in vegetation or slope or the presence of permeable or impermeable rock, could have on different graphs.

## Quick test ?

Without looking back at the last section, list the main items you should refer to when answering a question on a given hydrograph.

## Example

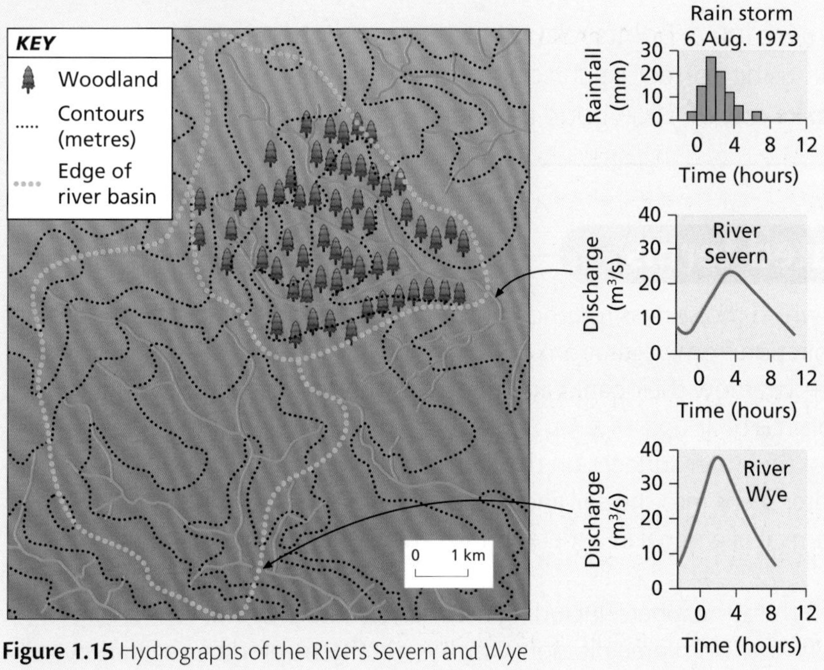

KEY
Woodland
Contours (metres)
Edge of river basin
0 1 km

Rain storm 6 Aug. 1973
Rainfall (mm)
Time (hours)
River Severn
Discharge ($m^3/s$)
Time (hours)
River Wye
Discharge ($m^3/s$)
Time (hours)

**Figure 1.15** Hydrographs of the Rivers Severn and Wye

Study Figure 1.15.

(a) Describe the differences in the hydrographs for the River Severn and the River Wye after the storm of 6 August 1973.

(b) Suggest reasons for these differences **10 marks**

## Sample answer

*The river Severn basin is covered by woodland. This would delay the water from reaching the river basin as quickly as the water in the Wye basin as there is less vegetation (✓). Some of the rainfall would be absorbed by the leaves of the trees and plants (✓) and also some would be taken up by the roots (✓). This accounts for the longer delay time at the beginning of the river Severn's hydrograph. (✓) The vegetation is also responsible for the lower height reached by the Severn – 23 m3/s compared to the high 47 m3/s in the Wye basin. (✓) The River Severn's drainage basin is also situated on gentler slopes than the Wye. This factor also reduces the run-off in the Severn's basin. (✓) The river Wye takes a shorter period of time, however, to return to its normal discharge level – about 9 hours. (✓) This is due to the water being held back longer in the Severn's basin.*

## Comment and marks

This candidate has chosen to answer both parts together, which is acceptable.

The first part of the answer correctly compares both basins in terms of the rate of discharge and relates this to the presence of woodland in the River Severn basin, thus gaining marks.

A further mark is obtained by relating the impact of woodland to the delay time in the River Severn's hydrograph.

The answer makes references to the discharge rates using data from the graphs. This obtains further marks.

There are no marks for the statement on the slopes of the Severn basin since the contour pattern on the map does not confirm this.

A final mark is obtained for the descriptive points relating to the shorter time for the Wye to return to its normal discharge, again using correct calculations from the graphs.

In total, the answer merits **7 marks out of 10**. A good answer.

## Key words and associated terms

**Evapotranspiration:** The process by which moisture is returned to the atmosphere by direct means through evaporation combined with transpiration from vegetation.

**Floodplains:** Areas formed when rivers overflow their banks and flood surrounding flat land, leaving deposits of silt, until eventually a flat floodplain is built up.

**Hydrosphere:** The name given to all of the water surfaces on the Earth.

**Infiltration:** The process by which water seeps into the soil and subsoil.

**Lag time:** A time delay between the arrival of a signal in a meteorological measuring instrument and the response of that instrument.

**Precipitation:** All forms of moisture in the atmosphere, including rain, hail, sleet and snow.

**Quick flow:** The surface movement of water from precipitation, which is not interrupted by vegetation and that runs as a shallow, unchannelled sheet across the soil.

Chapter 1.3

# Lithosphere

## Key point !

The lithosphere is the name given to the variety of physical landscapes, specifically (for the purposes of the Higher Geography exam and course) those relating to the United Kingdom, and the processes that created them. The types of landscapes that you are required to know about include glaciated landscapes and coastal landscapes.

## Glaciated landscapes

During the last 2.5 million years the British Isles have at different times been covered with large sheets of ice. These periods are called **glaciations**, and there may have been up to twenty different glaciations during the period known as the Ice Age.

Ice advanced southwards as the climate became colder and retreated as it became warmer. These advances and retreats are termed **glacial** and **interglacial periods** respectively.

### Key point

You are expected to know the main features of physical landscapes and the processes involved in their formation.

### Glacial formation processes

- Glacial **erosion** occurs through two processes:
  - **Abrasion**: a sand-papering effect as the ice moves across the land.
  - **Plucking**: pieces of rock are torn away from the land. Abrasion produces smoothed surfaces and plucking tends to produce jagged features.
- The **glaciers** consist of ice in which melting ice near the base of the glacier causes a process within the glacier known as **internal deformation**.
- Water gets trapped in cracks in rocks and alternately freezes and thaws, causing the rocks to break up. This is called '**freeze–thaw**' or '**frost shattering**'.
- The weight of the ice causes the glacier to slide on top of the melting ice. This is called **basal sliding**.
- The process of melting is called **ablation**.
- The rate of flow of the glacier depends on the type of rock over which it flows, the amount of ice in the glacier and the slope of the land.
- As glaciers move they erode and deposit material at their margins; namely, the front and sides. Melt water flowing from the glaciers further erodes and deposits material in processes that are described as **fluvioglacial**.

# Formation of erosional features in glaciated landscapes

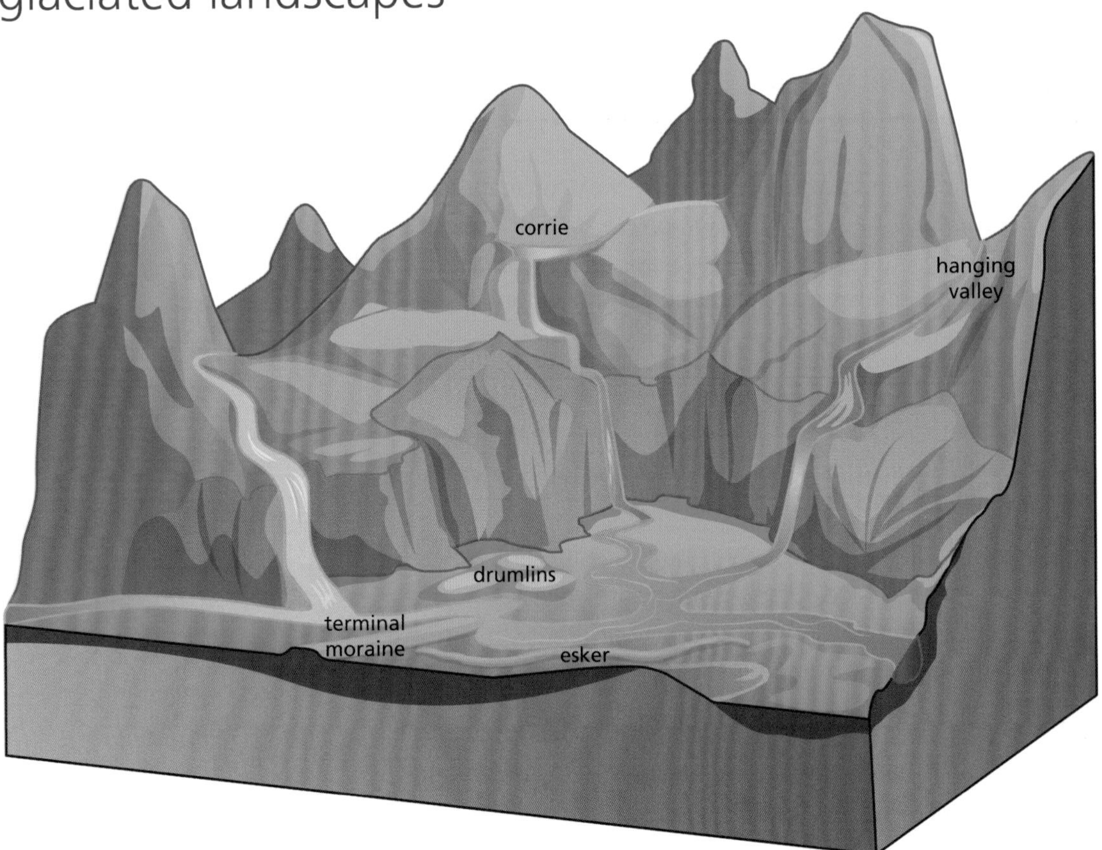


**Figure 1.16** A landscape of a glaciated upland and lowland glacial depositional features

1. **U-shaped valleys:** As a glacier moved downhill through a valley, the shape of the valley was transformed. Material called **boulder clay** was deposited on the floor of the valley. As the ice melted and retreated, the valley was left with very steep sides and a wide flat floor. A river or stream may have flowed through the valley due to melt water from the glacier. This replaced the original stream or river and is termed a **misfit stream**.

   The material that was pushed in front of the glacier, and left as the glacier melted, is called **terminal moraine**. If the valley dammed by the **moraine** flooded a lake was created, which may have twisted and turned and therefore is termed a **ribbon lake**.
2. **Hanging valleys:** The sides of the U-shaped valley are usually high and steep. During the Ice Age, tributary valleys often had smaller glaciers. The glacier in the main valley cut off the bottom slope of the tributary valley, leaving it high above the main valley. Tributaries of the main valley will therefore plunge from the slopes of the main valley into the bottom of the valley. These smaller valleys are called hanging valleys.
3. **Corries, arêtes and pyramidal peaks: Corries** are steep-sided hollows in the sides of mountains where snow has accumulated and gradually compacted into ice. The rotational movement of ice in the hollow causes considerable erosion, both on the floor and on the sides of the depression. The erosion on the floor is caused by abrasion, the floor becomes concave in shape and the edge takes on a ridge-shaped

*Key point* 

Erosion by glaciers produced several postglacial landforms, which you should be able to identify and explain how were formed.

appearance. At the sides, plucking of rocks takes place as the ice moves forward and the back wall of the depression becomes very steep. As the corrie fills up with ice, eventually it cannot contain any more and some of it moves down the slope to a lower level. This is the beginning of a glacier. Occasionally, as the ice melts, melt water fills the corrie and forms a corrie lake, called a **corrie loch** in Scotland and **tarn** in England.

**Arêtes:** Often, corries developed on adjacent sides of a mountain and when they were fully formed they were separated by a knife-shaped ridge termed an arête.

**Pyramidal peaks:** If corries develop on all sides of a mountain, the arêtes will form a jagged peak at the top. This feature is called a pyramidal peak. These are further sharpened by frost action.

**Quick test**

Write down a list of the main features of an upland glaciated area.

## Formation of depositional features in lowland glaciated landscapes

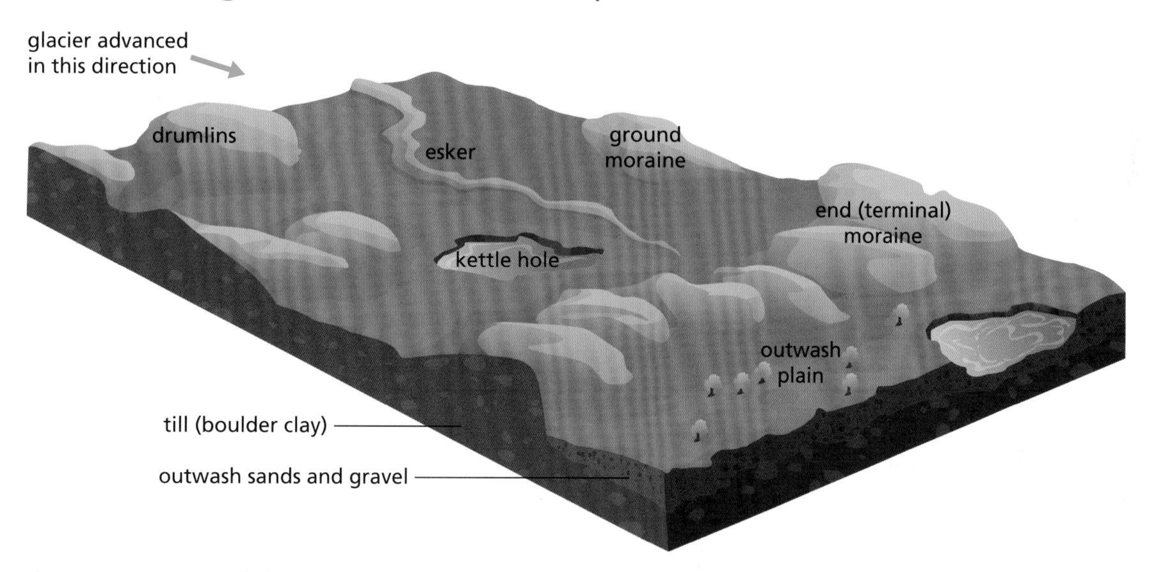


**Figure 1.17** Features of a lowland glaciated landscape

1. **Moraines:** Moraine consists of material known as **boulder clay and till**, which has been eroded and transported and deposited by the glacier. This material may be dumped at the end or snout of the glacier and is called **terminal moraine**.

   Those formed by material dumped at the sides or in the middle where two glaciers came together are called **lateral** and **medial moraines** respectively.
2. **Erratics:** These are large boulders that have been lifted, carried and deposited by the glaciers some distance away in a different part of the country. The rock type of the erratic is usually different from the rocks that are common to the area in which it is deposited.
3. **Outwash plains:** These are gently sloping plains consisting of sands and gravel. These have been deposited by melt water streams flowing out from the ice sheet and carrying material collected by the glacier.
4. **Eskers and drumlins: Eskers** are elongated ridges of coarse, stratified, fluvioglacial sands and gravels and are thought to have been formed by melt water tunnels within the lower parts of the glacier that deposited the material.

**Drumlins** are oval-shaped mounds that can be up to 100 metres high and have a 'basket of eggs' look to them. The material in them was deposited due to friction between the ice and the underlying rock causing the glacier to drop its load.

5 **Ribbon lakes:** These are large, narrow lakes occupying a U-shaped valley. When a glacier moves along its valley, changes in the rate of flow caused by extension or compression may lead to increased deepening of sections of the valley floor. Less resistant rock is more deeply eroded by the glacier and when the glacier retreats the deepened sections fill with melt water. Sometimes moraine creates a wall across a valley and a ribbon lake then fills the valley behind it. These lakes remain long after glaciation has ended, supplied by rainfall and subsequent streams and rivers.

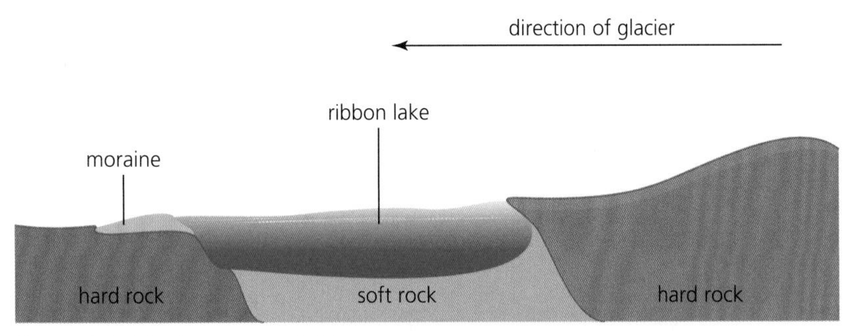


**Figure 1.18** The formation of ribbon lakes

# Geographical methods and techniques

Figure 1.19 gives an indication of what the features of an upland glaciated landscape will look like on an Ordnance Survey map. Study these and try to identify similar patterns on different map extracts.

In the examination you may be asked to identify features of an upland glaciated landscape from an Ordnance Survey map and to explain how selected features were formed.

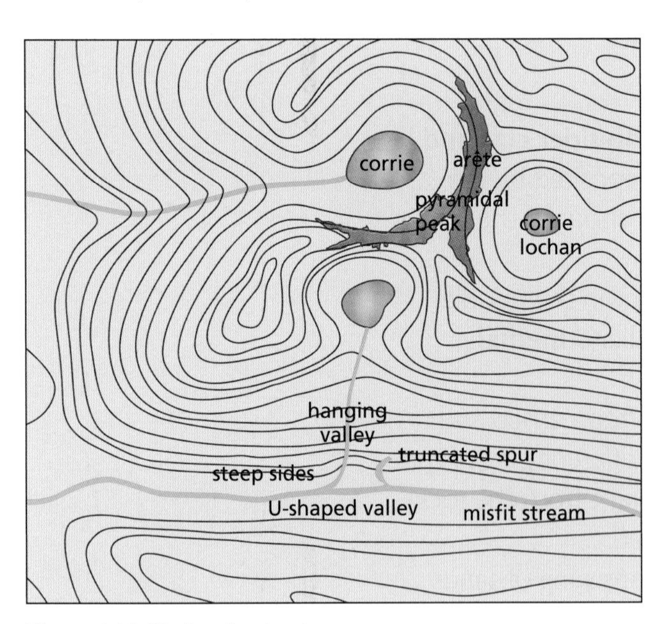


**Figure 1.19** Glaciated uplands

As well as identifying the features from the maps, you will probably also be required to name examples of these features and to provide appropriate six-figure grid references.

# Coastal landscapes

You should be able to identify physical features of a coastal landscape and explain how they were formed, including: wave-cut platforms, caves, arches, stacks, headlands, bays, spits, bars and tombolo. Some of these features are shown in Figure 1.20.

You should be able to identify physical features of a coastal landscape and explain how they were formed.

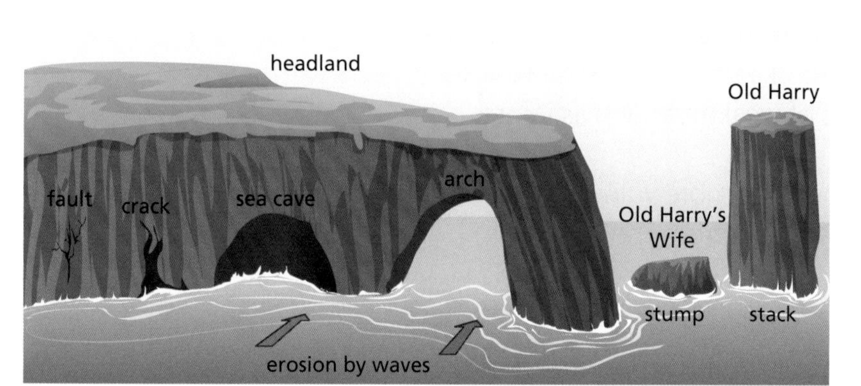


**Figure 1.20** A typical coastal landscape with headlands, caves, arches and stacks

You may be asked to identify some of these features from an Ordnance Survey map of a coastal area. These are shown in Figure 1.21.

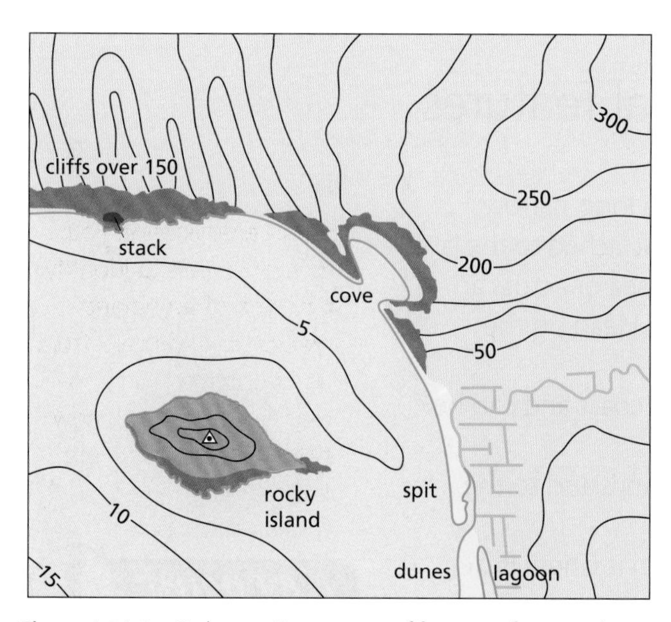


**Figure 1.21** An Ordnance Survey map of features of a coastal area

## Formation of erosional coastal features

**Cliffs, wave-cut platforms, headlands and bays:**

- **Cliffs** are formed by wave action undercutting land that meets the sea. This occurs at about high tide level. A notch is cut and, as the land recedes, the cliff base is deepened by wave erosion. At the same time the cliff face is continually attacked by weathering processes, and mass wasting such as slumping occurs, causing the cliff face to become less steep.

- When high, steep waves break at the bottom of a cliff, the cliff is undercut and forms a feature called a **wave-cut notch**. Continual undercutting causes the cliff to eventually collapse and as this process is repeated the cliff retreats, leaving a gently sloping **wave-cut platform**.
- The slope angle of the wave-cut platform is less than 4°.
- When resistant rocks alternate with less resistant rocks along a coast and are under wave attack, the resistant rocks form **headlands** while the less resistant rock is worn away to form **bays**.
- Bays and headlands can also develop in a single rock structure, for example limestone, which has lines of weakness such as joints or faults.
- Although the headlands gradually become more vulnerable to erosion, nevertheless they protect the adjacent bays from the effects of destructive waves.

**Caves, arches and stacks:**

- **Caves** are formed when waves attack cliffs with resistant rock along lines of weakness such as faults and joints.
- The waves undercut part of the cliff and can cut right through the cave to form an **arch**.
- Continual erosion causes the arch to widen and eventually the roof of the arch collapses to leave a piece of rock left standing, called a **stack**.

## Features of depositional coastal features

**Spits and bars:**

- **Spits** result from marine deposition and consist of a long, narrow accumulation of sand or shingle with one part still attached to the land.
- The other end projects at a narrow angle into the sea or across an estuary. This end is often hooked or curved. Spurn Head on the Humber estuary is a good example of a spit.
- **Bars** are ridges of sediment formed parallel to the coast and can be exposed at high or low tides.
- Often bars form barriers across bays. If a bar joins an island to the mainland it is called a **tombolo**.
- A very well-known example of a tombolo in Britain is Chesil Beach near Weymouth.

## Coastal modification processes

- **Corrasion** (abrasion) is caused by waves throwing beach material against cliffs.
- **Attrition** happens when waves cause rocks and boulders to break up into small particles by bumping them together.
- **Hydraulic action** is the process of waves compressing air in cracks in cliffs. As a result, any weaknesses in the rock face of cliffs and headlands are widened and create caves.
- **Longshore drift** occurs when waves remove material from beaches and deposit this material further along the coast.

*Key point* !

You should be able to describe and explain the processes involved in the modification of coastal landforms, including attrition, corrasion, hydraulic action, longshore drift, sea level changes, slumping, rock falls and cliff-line retreat.

- **Sea level changes** happened in postglacial periods when large amounts of ice melted and caused sea levels to rise, often drowning parts of coastlines.
- **Slumping** is the movement of surface rocks or superficial material that has become detached from a hillside or cliff face.
- **Rock falls** occur when small blocks of rock become detached from a cliff face due to the sea undercutting the cliff along joint patterns.
- **Cliff-line retreat** happens when the cliff face is gradually worn back by slumping, undercutting and rock falls.

**Example**

Explain how a corrie is formed.

You may wish to use an annotated diagram(s). **10 marks**

## Sample answer

A corrie begins when snow collects in a north facing hollow. (✓) If the weather is cold throughout the year then the snow doesn't melt and as it continues to snow, the air gets squashed out of the snow turning it into neve. (✓) This can happen in several different areas. The ice starts to move out of the hollow due to its weight and the force of gravity. (✓) Processes like plucking take place. (✓) This is when ice freezes onto the bare rock and pulls loose bits away from the sides and back making them steeper. (✓) Abrasion takes place at the base of the corrie (✓) as rocks incorporated in the glacier act like sandpaper and deepen the hollow. (✓) Frost shattering steepens the back wall (✓) as rock fragments break off due to the constant expansion and contraction of the rocks, due to temperature changes. (✓) The glacier starts to rotate and move out of the corrie leaving a lip. (✓)

## Comments and marks

This is an excellent answer. It explains the formation in a logical order, from its beginning in the hillside to its formation. Processes like plucking gain marks, but the candidate gains additional marks by explaining the processes. This answer would achieve **10 marks out of 10**.

## Key words and associated terms

### Physical landscapes

**Abrasion:** The process by which rocks within ice sheets and rivers scrape and erode the land over which they pass.

**Arête:** A narrow ridge between two corries, formed as corries are formed on two adjacent sides of a mountain.

**Attrition:** An erosional process that takes place in both rivers and at the coast. At the coast, attrition happens when rocks and pebbles carried by the waves smash into each other, wearing each other away and gradually becoming smaller, rounder and smoother.

**Corrie:** An armchair-shaped hollow on the side of a mountain, formed by ice filling a hollow and eroding the side of the mountain by abrasion and plucking, and by rotational movement at the base of the hollow.

**Drift:** Material deposited by a glacier, made up of two main parts: 'Till' deposited under the glacier and 'outwash' formed by melt water streams carrying particles of material from the debris under the glacier.

**Drumlin:** An oval-shaped hill formed from deposits within a glacier.

**Erosion:** The process by which rocks and landscapes are worn away by agents such as moving ice, wind, flowing water and sea/wave action.

**Erratics:** Rocks or boulders that have been moved by ice sheets from their original location and left in other parts of the country during the Ice Age.

**Eskers:** Long ridges of sand and gravel deposited by rivers that flowed under ice sheets.

**Freeze–thaw action:** When water trapped in cracks in rocks alternately freezes and thaws, causing the rock to break up.

**Frost shattering:** Similar to freeze–thaw; caused by water turning to ice and expanding as it melts to put pressure on the rock, eventually causing it to shatter.

**Glacier:** A large mass of moving ice that changes the shape of the land over which it is passing.

**Hanging valley:** A valley that is situated on the slopes of a U-shaped valley and a tributary of the main valley.

**Misfit stream:** A U-shaped valley is usually occupied by a small river that is known as a 'misfit stream' since it was not the original river flowing through the valley.

**Moraine:** Material that is deposited by glaciers. Different types include 'end moraines'/'terminal moraines' formed at the front of the glacier as it melts; 'lateral moraines' formed at the sides of glaciers; and 'medial moraines' formed in the middle of glaciers or at the edges where two glaciers meet.

**Plucking:** The process by which moving ice tears rocks from the surface over which it moves.

For associated coastal terms, refer to the section on coastal features (pages 25–27).

# Soils: formation, properties and soil types

**Key point**

To answer questions in this chapter you have to be able to show knowledge of the properties and formation processes of soils.
It is important to understand the processes involved in soil formation and the main characteristics of a soil profile.

## Soil formation

Soils are formed from the processes of weathering, erosion and deposition from **parent rocks**.

The type of soil present is a result of a long formation process involving a range of elements including:

- inputs such as weathered rock
- the decomposition of dead vegetation into organic matter
- the impact of bacteria derived from decayed matter
- the influence of heat (solar energy)
- water movement, either downwards or upwards
- the input from other forms of life, such as insects, worms and other animals.

**Key point**

You should be able to recognise and describe the properties of soils from a soil profile.

**Key point**

You may also be asked to draw a soil profile!

## Properties

Soils consist of several layers called **horizons** or **soil profiles**.

Figure 1.22 shows a basic model horizon or soil profile.

- The layer closest to the surface is called the **Ao** layer, and it contains the organic material derived from dead plants and other organisms. This is known as humus. The depth of this layer varies greatly from soil type to soil type. The humus forms the lowest section of the Ao horizon.
- Below the Ao horizon lies the **A** horizon, which is the soil layer proper. This consists of a mixture of humus and other mineral particles.
- A further layer, known as the **B** horizon, is found beneath the A horizon and consists of coarser material. This is the subsoil layer.
- There are two further layers, called the **C** horizon, consisting of weathered rock fragments, and the bottom layer or **D** horizon, which is the parent rock layer.
- Depending on the variations within these different horizons, it is possible to group soils into a classification system known as **zonal soils**. These soils have certain common characteristics and are closely linked with the climatic and vegetation zones in which they are found.

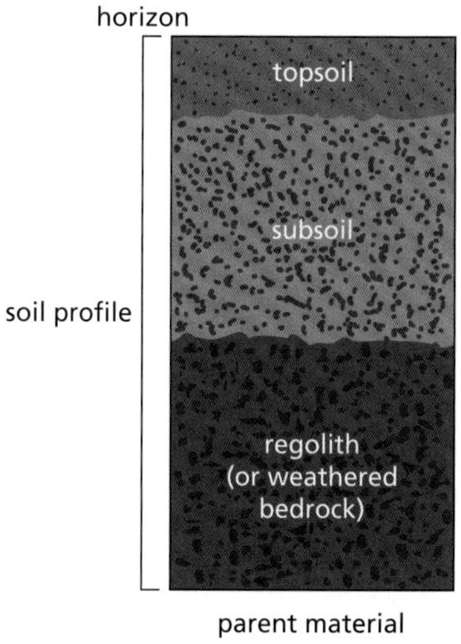

horizon
topsoil
subsoil
soil profile
regolith (or weathered bedrock)
parent material

**Figure 1.22** A basic soil profile

# Processes

Various processes are involved in the formation of soils in general. These processes depend on factors such as climate, relief, organisms, parent material and time.

- The first stage of the process involves the weathering of parent rock over a considerable period of time.
- The next stage of formation is the result of adding water, gases, living organisms and decayed organic matter.
- The rate of weathering of the parent rock greatly depends on the climate. The quickest rate of breakdown occurs in hot, humid climates.
- If rainfall is heavy, water moving downwards through the soil transports minerals downwards in a process called **leaching**. Leaching occurs where there is excessive rainfall.
- Leached soils such as podsols tend to be acidic, whereas **capillary action** tends to produce more alkaline soils. Capillary action is the process by which soil moisture moves through the fine pores of the soil.

The type of vegetation in an area is related to the amount of precipitation. This vegetation provides humus, and more humus is found in tropical forests than in tundra areas.

Two other processes are important; namely, **eluviation** and **illuviation**.

- Eluviation is the washing out of material, that is, the removal of minerals, such as calcium and aluminium, and organic material from the A horizon.
- Illuviation is the deposition of this washed material in the subsoil. Heavy minerals are deposited much deeper, and this affects the fertility of the soil.
- Organisms within the soil affect the breakdown and decay of vegetation and therefore impact on the depth of the humus layer.
- Any living creatures, such as insects and worms, affect the development of the soil since they can aerate and expose the soil to air and can add to the chemical balance of the soil through their excreta.
- In areas with cool climates and mainly coniferous forest, and where precipitation greatly exceeds evapotranspiration, a process known as **podsolisation** is common:
  - Percolating rainwater becomes quite acidic as it passes through an acidic humus formed from falling pine cones and needles.
  - This water dissolves and removes iron and aluminium oxides from the topsoil and leaves behind a high level of silica in the A horizon, which is bleached and drained of coloured minerals.
- When waterlogged conditions exist in the soil due to the loss of water from the soil being restricted (if, for example, the subsoil is full of stagnating water, which loses oxygen), a process known as **gleying** occurs. This often happens in poorly drained areas where the land is more gently sloping.

# Soil types

Figure 1.23 illustrates the profiles of podsols, brown earth soils and gley (tundra) soils.

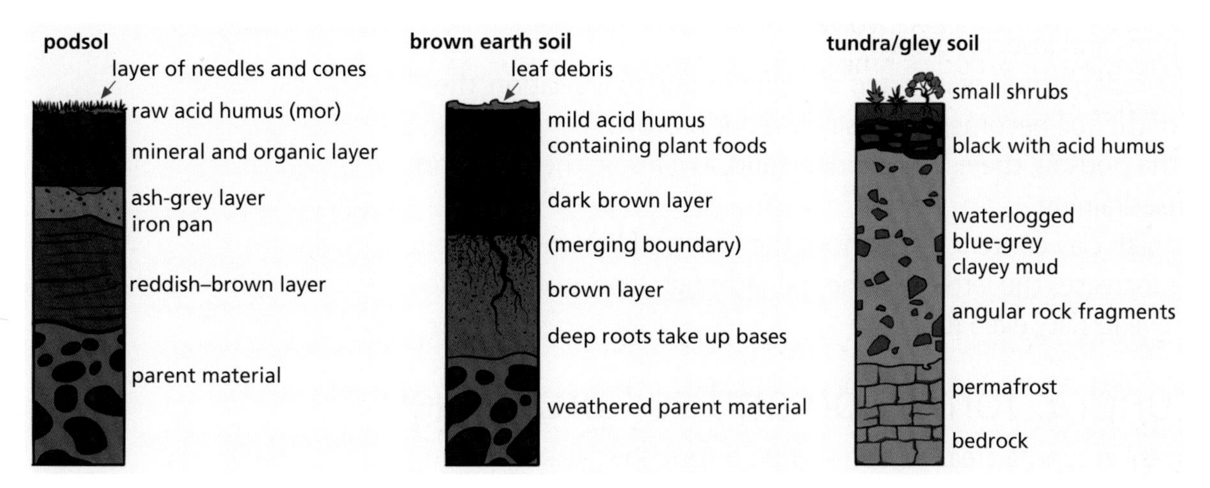


**Figure 1.23** Selected soil profiles (podsols, brown earth and tundra/gley)

## Podsols: formation and soil properties

- These soils are found in a wide belt across the northern hemisphere, particularly in areas of taiga or coniferous forests.
- Falling pine needles and cones create an acidic humus called **mor**.
- The soil has well-defined layers since there is very little movement and mixing of the horizons due to the absence of earthworms because of the cold conditions.
- The upper A horizon has an ash-grey colour due to the removal of the minerals.
- With aluminium and iron oxides concentrated in the B horizon, a cementing effect takes place between the A and B horizons, forming a hardpan that seriously affects the drainage through the soil. This results in the upper layers becoming waterlogged.
- The B horizon is reddish-brown in colour due to the iron oxides.
- The subsoil consists of weathered parent rock.
- Decomposition of the Ao horizon is very slow due to the cold climatic conditions.
- In addition, moderate precipitation and melt water released from snow or ice during the spring produces leaching of iron and aluminium oxides from the A horizon, leaving a high silica residue.

*Key point* 

You should be able to describe and explain the effects of climate, relief and drainage on the formation and properties of the following soils: podsols, brown earth soils and gley/tundra soils.

## Brown earth soils: formation and soil properties

- These soils are associated with areas of deciduous forest. They are sometimes referred to as **alfisols**.
- The humus layer is thick and generally fertile due to the variety of vegetation that is decayed. With warmer conditions, the leaf litter that accumulates in autumn decomposes quickly due to organisms in the soil.

- The humus is less acidic and is referred to as **mull**.
- Precipitation exceeds evaporation sufficiently to cause leaching. However, there is not enough to cause podsolisation. There is an absence of bases, especially calcium and magnesium, in the A horizon.
- The horizons merge more than in podsols due to the activity of earthworms and insects (biota).
- With the redeposition of iron and aluminium due to illuviation, the colour of the soil becomes increasingly reddish-brown.
- Unlike the podsols, there is no hardpan and, as a result, the soils tend to be free draining.
- There is high clay content throughout the profile of brown earth soils and this increases the fertility of the soil, although lime is often added to improve fertility even further.

## Gley/tundra: formation and soil properties

- The subsoil in tundra areas remains frozen throughout year. During the brief summer the ground surface thaws but the melt water cannot drain freely due to the frozen subsoil or permafrost layer. This results in the soil becoming waterlogged or gleyed.
- Bacterial action is very restricted due to the cold temperatures. The waterlogged soil lacks oxygen and the tundra vegetation is also very limited.
- Alternate periods of freezing and thawing cause great disturbance and mixing of the soil. Consequently the horizons are less well defined.
- The A horizon contains black, acidic humus, which is only partially decayed due to low temperatures.
- The B horizon is bluish-grey in colour, with clayey mud.
- Fragments of weathered parent material are often found within the B horizon.
- The cold conditions of the climate severely restrict the use of these soils.

**Key point !**

You should be able to describe and analyse soil profiles, particularly those of podsols, brown earth and gley/tundra profiles.

## Geographical methods and techniques

### Analysis

- You should begin your analysis by noting the various constituents of the different horizons within the profile, beginning with the topsoil.
- Describe the thickness of this layer and refer to reasons for this, for example the depth of the humus layer and vegetation.
- You should then discuss each layer in turn, referring to thickness, content, colour, texture, water content and whether the layers of the A, B and C horizons are well defined or whether mixing has occurred.
- You may explain the properties of the horizons by referring to factors such as climate, vegetation, processes of leaching, eluviation and illuviation and soil biota.
- From this description and explanation you should be able to deduce the type of soil that the given soil profile shows.

**Quick test**

Draw a soil profile for one of the following soil types:

- podsol
- brown earth soil
- gley/tundra soil.

## Example

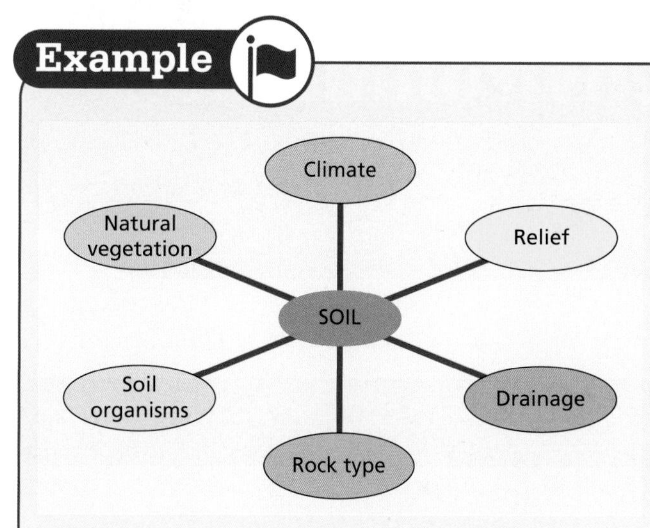


**Figure 1.24** Main factors affecting soil formation

Study the diagram above and explain how brown earth soils are formed. **8 marks**

### Sample answer

*Brown earth soils are found in temperate areas of the world. The thick humus layer is formed from decaying leaves, for example from deciduous trees such as chestnut and oak (✓) which shed their leaves in the autumn. (✓) Organic material decays well in the warm climate thus adding humus to the soil. (✓) Long tree roots go deep into the soil allowing minerals to be moved about. (✓) There is a little leaching of the soil. Leaching of the minerals can cause an iron pan to develop. (✓) The iron pan hinders soil drainage. (✓) Worms and moles help to mix the decaying litter which in turn helps to maintain fertility. (✓) Worms and moles also help to aerate the soil and therefore prevent the formation of distinct layers in the soil. (✓)*

### Comment and marks

This is a very good answer that explains the main processes in the formation of this type of soil. Marks are gained from statements on the formation of a humus layer, the decay of organic material in a warm climate, the leaching process and its impact on drainage and finally the impact of animal life on the soil.

The answer merits a total of **8 marks out of 8**.

## Key words and associated terms

**Capillary action:** The upward movement of water in a soil.
**Eluviation:** Water percolating through soil.
**Gleying:** When soils become waterlogged with stagnant water that is unable to drain away, the entry of oxygen into the soil is restricted, turning the red iron oxides into blue-grey iron oxide.
**Illuviation:** The process by which material is deposited in the subsoil.
**Leaching:** The process by which minerals are carried through the soil by percolating water.
**Parent rock:** Rocks from which particles are worn away to eventually become other rocks.
**Soil profile:** A diagram that shows the different layers or horizons within a soil.

# Section 2 Human environments

## Chapter 2.1
## Population

# Methods and problems of data collection

## National censuses

Data on a country's population is collected through the process known as a national **census**. In the UK a national census is undertaken every ten years. By law, one person in each household is required to complete a census form for that household.

It is similar for other premises such as hotels, hospitals and residential homes. Similar censuses are carried out in **developed countries** and **developing countries**.

The census form contains a wide variety of questions relating to details such as date of birth, gender, employment status, ethnic origin and residence.

The costs of carrying out regular and accurate censuses are very high.

The census in the UK employs people to distribute census forms and collect them after a few days. These people are called **enumerators**.

Once collected, the forms are analysed by full-time government analysts and the information is published and made available to the general public.

The information is used by the government to plan for the future, for example it is used to plan housing policies, education and transport and in the analysis of **population structures**.

The Office of National Statistics (ONS) carried out a consultation in 2013 to find out how future population data should be gathered. It concluded data could still be collected from using vital registration as well as compulsory yearly surveys. It also decided that the census from 2021 would be mainly online, replacing the traditional paper-based ten-year census. However, the ONS is aware that care will need to be taken to support those who do not have access to the internet and so will be unable to complete the census online.

*Key point* 

You should know how information about populations is gathered through a census, and that developing countries have more difficulties in carrying out a census than developed countries and that the data is less reliable.

## Censuses in developing countries

Developing countries such as Pakistan and Malawi have more difficulties in carrying out a census than developed countries and the data is less reliable. Reasons for this include:

- In countries where people move around a great deal, for example nomadic herdsmen, it is very difficult to keep track of populations.
- If the standard of education is poor, the population may be unable to read and complete the forms.
- Conflict such as wars and natural disasters, for example famine and floods, also makes it extremely difficult to carry out an accurate census.
- Some governments may have political reasons for not having accurate census details, particularly if there are issues related to ethnic minorities.
- Other reasons include, for example, the size of the country and difficulty with the terrain, such as in mountainous desert areas and rainforest areas.

In addition to a national census, other sources of information on populations include:

- Information from registry offices on births, deaths and marriages.
- National insurance records providing employment figures.
- Details from vehicle licensing departments on the number and type of vehicles currently in use.
- Records from all sectors of education including schools, colleges and universities on the numbers of people in full-time education.
- Information from local governments on local building developments, including council and private projects.
- Details of crime rates, prison occupation and figures on different categories of crime from police and other authorities that deal with this issue.
- Currently most of this information is held on computer and can be analysed by government statistic analysis staff to help governments plan for future needs.

# Population structure

You are often asked in exam questions to interpret population pyramids for developed and developing countries and account for the different structures. How do you do this?

Note that the background information on population structures should help to enhance the quality of your answer on 'consequences'.

# Population structure of a developed country

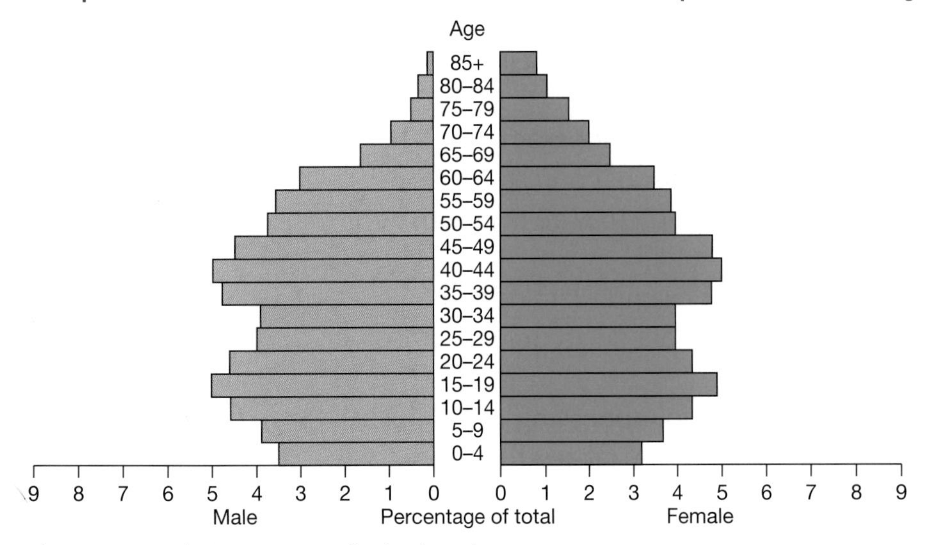


**Figure 2.1** Population structure of a developed country

## Main features

Figure 2.1 indicates the population structure of a developed country.

When asked about the main features you should mention:

- a fairly low **birth rate** in both males and females
- a bulge in the middle age groups, particularly 15 to 60 years, which shows that most of the population are within these age groups
- a fairly high percentage of the population within the upper age groups from age 60 upwards, showing high **life expectancy** and a population that has more older than younger people.

### Main reasons for the structure

The main reasons for this structure include:

- the widespread use of artificial birth control
- changes in the status of women, with many women having jobs and careers in preference to marrying young and starting a family
- the number of children per family usually being low due to widespread use of contraception and couples having children much later in their marriages
- the overall wealth of the country and the high **standard of living** enjoyed by its population
- the change in attitudes of younger people towards marriage and size of families and the opportunities for improving standards of living through having fewer children
- high standards of health care, education, housing and employment
- average income per head of population being high
- people living longer due to high standards of health care.

## Consequences

The consequences of this structure for a developed country include:

- fewer young people, which deprives the country of a suitable workforce for future generations

- more older people being cared for due to increased life expectancy
- underpopulation – where the birth rates and **death rates** are very low and are almost the same, population growth is very slow and in some cases is decreasing; when this happens the population structure becomes imbalanced
- increased demand for medical care and spending on care for elderly people, for example retirement homes
- a smaller group in the economically active age group, which has to support an increasingly economically dependent age group, for example through increased taxes to pay for health and social services.

# Population structure of a developing country

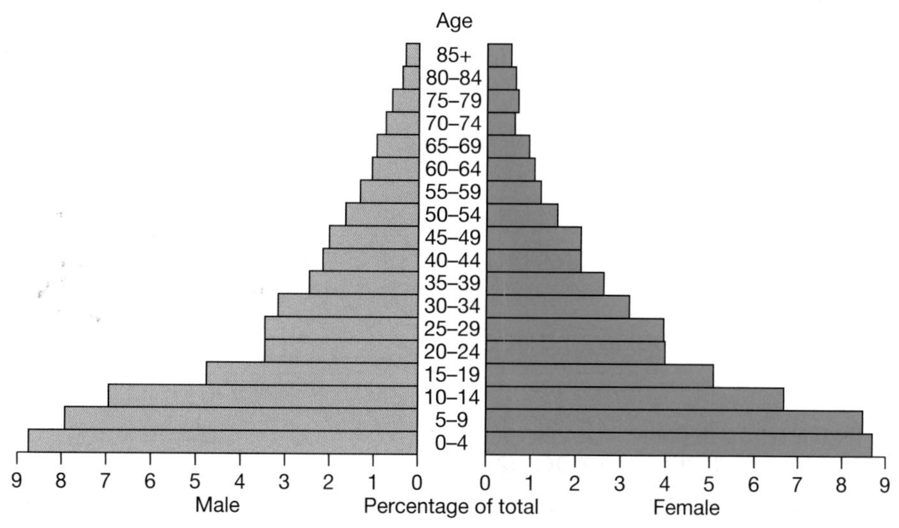


**Figure 2.2** Population structure of a developing country

## Main features

Figure 2.2 indicates the population structure of a developing country. Its main features include:

- a high birth rate in both males and females
- a large proportion of the population between the ages of 0 and 15 years
- a more definite pyramidal shape
- the numbers of people in the upper age groups above 15 years decreasing fairly rapidly
- very few people in age groups above 60 years, meaning that average life expectancy is low.

### Main reasons for the structure

This population structure is due to:

- a lack of social and economic development of the country
- usually low standards of personal wealth, industrial growth, health, education, food supply, housing, health care and employment
- the birth rate and death rate both being high
- families having a large number of children because **infant mortality** rates are so high; children work and bring in more income for the family and, when grown, look after parents when they become elderly.

## Consequences

The consequences of population structures for a developing country's economy and people may include:

- **Overpopulation**, which is said to exist whenever a reduction in the existing population would result in an improvement in the quality of life for the remaining population.
- A **lack of sufficient food** to meet demand.
- **Inadequate housing** for the population, particularly in cities and towns.
- Vast numbers of people in developing countries are forced to live in very poor accommodation such as **shanty towns**.
- Shanty towns lack the **basic facilities** of sewage, electricity and water supply and consequently disease levels are high.
- **Unemployment** is high since there are far too many people for the jobs available.
- **Poverty** is widespread. This is made worse by the lack of government financial aid.
- **Lack of services**, such as health centres, hospitals, doctors, sewage systems, clean water supplies, schools and colleges, creates problems of poor health and education standards.
- **Literacy rates**, that is, the percentage of the population that can read and write, are usually quite low.

*Key point*

You should be able to discuss the consequences of population structure. These consequences are closely related to birth and death rates in developed and developing countries.

As well as differences in population structures between developed and developing countries, in many developing countries there are significant differences between urban and rural areas.

Although for the purpose of examination questions on this topic you do not necessarily need to know about these differences, it is important to note them at this point. This is because these differences can have a significant impact on other issues raised later, in the section that deals with changes in cities in developing countries.

**Quick test**

Draw a basic diagram to show the structure of a developed and a developing country.

*Hints & tips*

*Make a list of the main differences in the diagrams you have just drawn and memorise it.*

# The main differences between urban and rural population structures

- There are wide variations in the percentages of urban and rural populations in many countries throughout the world.
- In developing countries the percentage of the rural population is usually higher than the urban population.
- In developed countries the vast majority of the population lives in urban areas.
- In Europe and North America more than 85 per cent of the population is urbanised.

- In developed countries the balance between rural and urban population is changing due to people leaving the countryside and migrating to urban areas.
- Often the structure of the remaining population is left imbalanced since there may be a larger proportion of older people.
- During the late nineteenth and early twentieth centuries, millions of people left rural areas in Europe to seek new lives abroad. This was true of, for example, Italy, Ireland, Germany and Russia, where many people from these countries emigrated to the USA for social, economic and political reasons.
- In developing countries, many people leave the poorer rural areas and migrate to cities in the hope of finding employment, homes, better education opportunities, better health care and generally a higher standard of living. In many case these expectations are never realised.

# Death rates in developed and developing countries

## A comparison between death rates in developed and developing countries

In developed countries the majority of deaths (80 per cent) occur among the oldest age group (over 65), whereas this age group accounts for only 25 per cent of deaths in developing countries.

In developing countries the majority of deaths occur in lower age groups, particularly young children. In the developed world this accounts for a very small proportion of deaths (under 10 per cent). The number of deaths in the 15–64 age group is also much higher in developing countries than in developed countries.

You should be able to compare different patterns of death by age group between developed and developing countries and suggest consequences of the differences.

### Reasons for the differences in death rates

- There are differences in life expectancy and infant mortality rates as a result of different living standards.
- There is a poorer quality of health and hygiene care available to citizens in developing countries than in developed countries.
- A poor-quality diet and food supply in developing countries often leads to malnutrition and in extreme cases famine.
- There are wide differences in the levels of economic development, with a lack of investment in housing, health and general infrastructure in developing countries.
- Low standards of education due to a lack of schools, colleges and teaching staff lead to a higher death rate in developing countries.
- Poor housing standards, with shanty towns common to most cities in developing countries, lead to higher death rates there.
- A prevalence of infectious disease due to poor sanitation and a lack of medical drugs and health care leads to higher death rates in developing countries.
- Developed countries enjoy much higher levels of provision in all of these areas than do developing countries.

# Geographical methods and techniques

## Interpretation of population graphs

- Analysing different types of graphs is important in the study of population.
- The analysis of **population models** by reference to the different patterns of birth rates and death rates throughout the four different life stages could be used to assess your geographical skills in this topic area.
- This would involve being able to take each stage in turn and provide an analysis of whether the rates are rising or falling, how fast or slowly this is happening and what the net effect would be on population growth. Note that you could be expected to explain *why* these changes are occurring, as well as describing what is happening.
- In the external examination this question could be combined with a question requiring you to explain the patterns in the different stages and/or comment on the effects of these changes on the population.
- You would draw on your knowledge of population trends and patterns and the factors responsible to help you explain population changes and their effects. You should be able to relate these to countries you have studied.

Key point

You should be familiar with a variety of population graphs and be able to interpret them.

## Interpretation of population data

- This skill involves being able to use population data to explain trends.
- Questions on these kinds of diagrams may ask you to examine graphs, tables or maps and to draw conclusions, for example you may be given a table that shows various measurements of population for two or more countries, which may contain details on life expectancy, birth and death rates, infant mortality rates and other indicators such as medical provision.
- You may be asked to explain differences between the countries, for example life expectancy. You can do this by first describing the differences and then referring to other information to support your conclusions on varying levels of development between the countries.
- Similarly, if you are asked to use diagrams such as population pyramids to explain differences between countries, you should be able to identify patterns of birth and death rates and life expectancy rates and to match them appropriately to countries that are either developed or developing.
- Your knowledge of factors that contribute to levels of development should help you to explain both the structures and the reasons for them.

# Migration patterns

## Migration

- **Immigration** is the permanent inward movement of people from other parts of the world to a particular country. This normally results in an increase in population.
- **Emigration** is the permanent outward movement of population from a country to another part of the world. This would normally cause population size to decrease.

**Key point**

You should be able to explain the causes and impacts of forced and voluntary migration on donor and receiving countries.

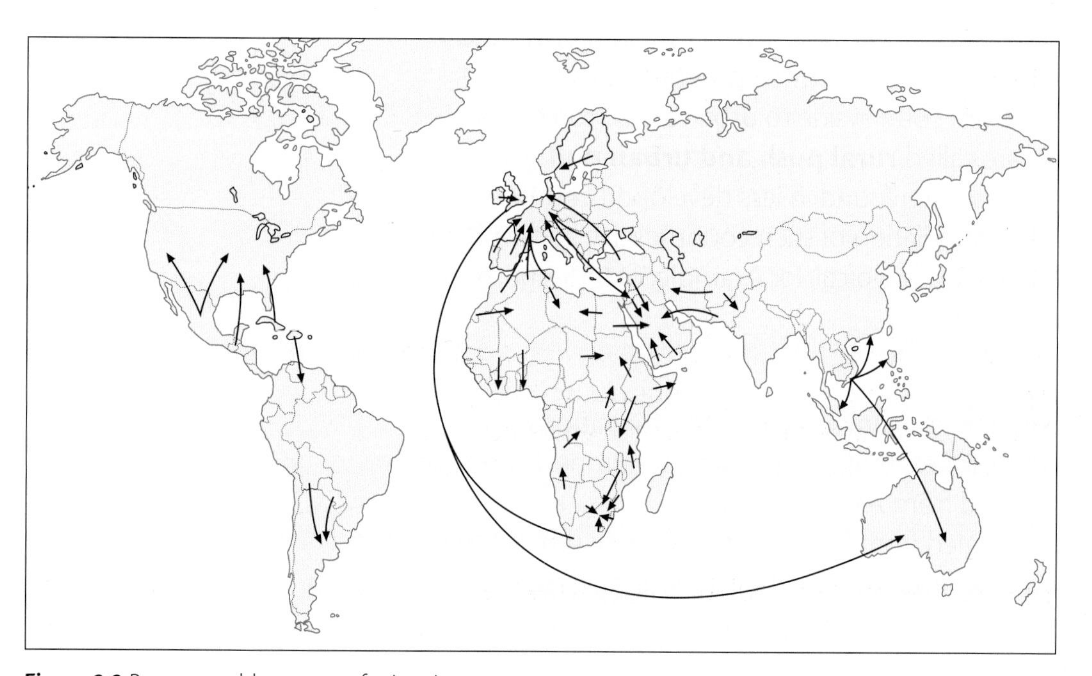

**Figure 2.3** Recent world patterns of migration

### Different types of migration include:

- **Voluntary:** when people choose of their own free will to move to another country.
- **Forced:** when a population has to move to another area against its wishes.
- **Long term:** when people leave their home area and live elsewhere for long periods of time.
- **Short term:** when people move to an area for a short period of time, for example a season.
- **Rural–urban:** when people leave the countryside to live in city areas.

### Reasons for migration

#### Voluntary migration

- People migrate from one area to another in the hope that they might improve their standard of living and general lifestyle.
- People often move from areas with high unemployment, low wages, poor housing, health and educational facilities, in the hope of improving on these conditions.

### Forced migration

- People may be forced to move to another region or country. Examples of this kind of migration include the African slave trade of the seventeenth and eighteenth centuries, during which up to 10 million Africans were forcibly moved from Africa to other parts of the world.
- Similar movements have occurred during periods of war, for example, during the First and Second World Wars large numbers of people were displaced from their homelands to other countries.
- Political and/or religious persecution.

### Rural–urban migration

- Conditions that cause emigration are known as **push factors** and those resulting in immigration are termed **pull factors**.
- When people move from countryside to urban areas (towns and cities) these factors are called **rural push and urban pull**. This pattern of movement is often found in less developed countries.
- A large portion of the population of such countries live and work in rural/agricultural areas. This is typical, for example, of India, Brazil and countries of south-east Asia.

### Illegal migration

- Many thousands of people are involved in illegal migrations throughout the world, for example from Africa to Europe, from South America to the USA and from Asia to Europe.
- These people are smuggled from their original country through secret routes to their country of destination. They have to pay large sums of money to illegal organisations that specialise in smuggling human traffic.
- Due to wars, persecution and natural disasters, many thousands of people throughout the world, including in Africa, the Middle East and Asia, have become refugees.
- They have been forced to migrate to other countries to obtain safety, aid and shelter and often live in huge refugee camps made from basic equipment and have only very basic facilities.

## Impact of migration

- Immigrants moving from rural to urban areas are often left disappointed and disillusioned.
- Many find themselves living in conditions that are worse than those they left behind.
- Many end up living in shanty towns within and on the outskirts of cities.
- Shanty towns have the poorest living conditions, including poor housing, lack of clean water supplies, no electricity and often the most basic systems of sanitation.
- Diseases such as cholera and typhoid are common in these areas and kill the weakest members of the community; namely, the poorest, the old and the very young.
- Immigration of different ethnic groups into countries can lead to problems of integration, racial tension and cultural differences.

- Some immigrants might become targets for abuse, and discrimination in terms of jobs and housing.
- Immigrants may be prepared to take lower-paid jobs and incur the anger of the existing population for doing so.
- Immigrants may also improve the cultural diversity and provide additional labour, especially in countries where the population is in decline due to falling birth rates.
- Ghettos may develop in inner-city areas involving groups of immigrants.
- Some countries pass legislation, both to limit immigration and to protect new immigrants.

## Advantages of migration

- Receiving countries acquire labour for a variety of occupations, many of which may be less popular and difficult to fill from their own populations, or for which they have a skill shortage. The home countries of the immigrants benefit from money sent back and from skills that have been acquired in the receiving country if they return to the home country. There is also reduced pressure on resources while they are away.
- Incoming immigrants may add to the depth of culture of the existing population in terms of customs, language and traditions.
- If the receiving country's population is in decline, the immigrant population may help to reverse the trend, providing more balance to the population structure.

**Key point !**

You should be able to discuss the advantages and disadvantages that migration brings to both the losing and receiving countries.

## Disadvantages of migration

- Emigration creates problems for the areas that lose people. In some cases, large numbers leaving result in serious depopulation.
- In small communities such as rural villages, for example in the highlands of Scotland and villages in southern Italy and rural India, the loss of young people results in a highly imbalanced population structure. There is a large number of old people left. Often within a few years the community dies out.
- Immigration into new areas, whether a country or a city, often results in serious problems due to the pressures created on the existing population.
- New immigrants may increase the demand for housing, employment, education and health care and will compete with the existing population for these services. In areas of relatively high unemployment, additional competition for employment, especially if the new immigrants are willing to work for less money than the established population, creates great resentment.
- Demand for accommodation causes problems, especially if the government is unable to supply the immigrants with housing.
- In developing countries, the main result of this has been the emergence of shanty towns.
- In developed countries, immigration from other countries can result in racial problems between ethnic communities, especially where language differences exist.

- Racial discrimination is common in some areas and shows itself in, for example, prejudice in employment and housing and open racial abuse.

## Strategies to manage migration

- In the European Union, migration is encouraged under European law. There is freedom of movement of labour.
- Many people from the poorer areas of the European Union leave their home country to seek employment in the wealthier countries of the Union.
- This causes many arguments between member countries.
- In other parts of the world, for example the USA, immigration is strictly controlled. Illegal immigration is a criminal offence.
- Anyone wishing to go to the USA to work must have a work permit called a **Green Card**.
- Barriers to migration include legislation in receiving countries that limits the number of immigrants permitted to enter the country.
- Legislation may include references to marital status, the ability to speak the language of the receiving country, professional or trade qualifications, political considerations or criminal records.
- Social barriers may include racial bias and problems with housing, education, employment and ethnic integration.

### Example

For a named voluntary migration you have studied, discuss the impact on the receiving country. **6 marks**

### Sample answer

*The advantages for Germany were that the Turks were willing to do the unskilled jobs that nobody else wanted at the time for a low wage. (✓) It has been said that without the Turks, Germany's hospitals, electricity and transport services would cease to work. Disadvantages are the fighting and social problems where the Turks don't mingle with the Germans (✓) and introduced their own culture into the city. (✓)*

### Comment and marks

The answer is credited for the references to the type of jobs taken by immigrants and that no one in the host country was willing to do them. No credit is gained for the mention of the low wages and that certain service industries depended on this labour as it is not developed sufficiently from the first sentence to gain a mark for a developed point. A second mark is obtained by the mention of the social problems that arise, although the last part, about culture, could be seen as an advantage rather than a disadvantage and would gain a mark.

A total of **3 marks out of 6** would be awarded.

## Key words and associated terms

**Birth rate:** The number of births per thousand of the population in any country in any year.

**Census:** A numerical count of the population, financed and carried out by the government at set periods of time, for example ten-year intervals.

**Death rate:** The number of deaths per thousand of the population of any country in any given year.

**Developed countries:** Countries that have a high standard of living or high physical quality of life.

**Developing countries:** Countries in which the population generally has a low standard of living.

**Infant mortality:** The number of children below the age of one year who die per thousand of the population.

**Life expectancy:** The average age a person can expect to live in any given country. This is a good indicator of the level of development since people in more developed countries tend to live longer due to better health care, better diets, higher standards of education and housing and so on.

**Population density:** The average number of people within a given area, for example 100 people per square kilometre.

**Population model:** A model that shows different stages of population growth based on the relationship between birth and death rates.

**Population structure:** The grouping of the population of a country by age and sex. Inspection of the structure may indicate trends in birth and death rates, life expectancy and the possible impact of factors such as war and migration on the population.

**Standard of living:** The level of economic well-being of people in a country.

# Chapter 2.2
# Rural

## Rural land degradation: deforestation

### Deforestation (rainforests)

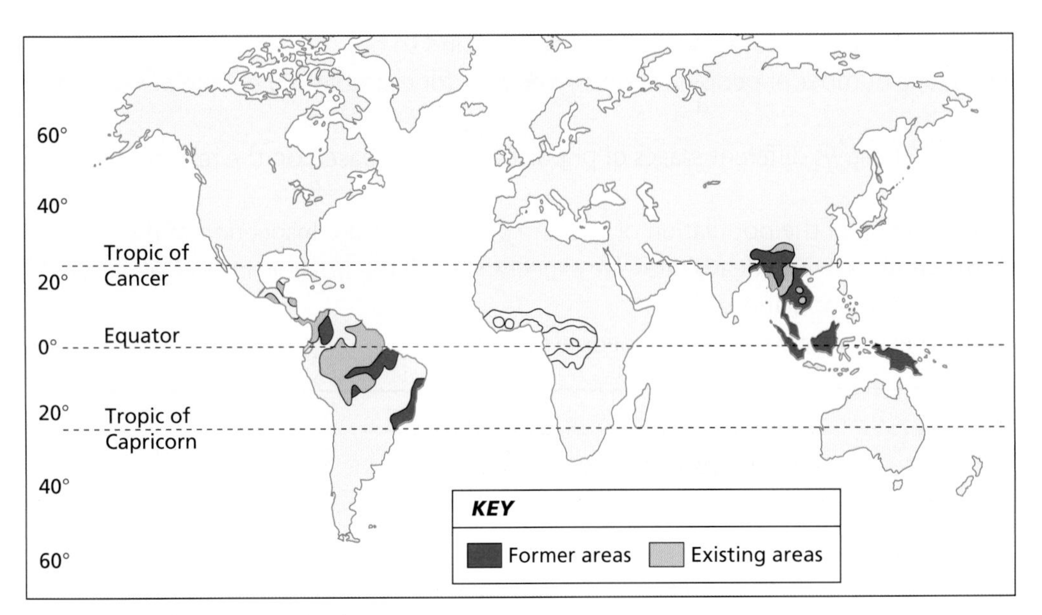


**Figure 2.4** World examples of deforestation

Rainforests are located in the land areas between the Tropics of Cancer and Capricorn in all continents.

#### Key point

For a rainforest area you have studied, you should be able to explain the impact of rural land degradation and how this is managed.

### Physical features

- The main features of the climate are high temperatures and high rainfall throughout the year.
- Vegetation consists mainly of hardwood trees such as teak and mahogany, rising to over 50 metres and forming a canopy of several layers.
- A wide variety of plant species and wildlife exists within the forest.
- Soils include forest soils that, when exposed to rainfall and Sun through the removal of trees, become leached. In this process, minerals rise through the soil to the surface and form what is termed a **hardpan**.
- When this happens the soil becomes infertile.

### Human features

- Population density is generally low within the forests and consists largely of indigenous tribespeople, communities of farmers, rubber tappers and miners.

- The main farming systems that exist include subsistence shifting cultivation, large-scale ranching and small-scale crop farming:
  - Shifting cultivation is the traditional type of farming practised by tribespeople. The system is based on small groups of native Indians who clear a small part of the forest and, over a period of about five to eight years, cultivate small patches of land. Crops include yams and manioc, and fish from local rivers supplements the diet, along with meat from animals hunted in the forest and local forest fruits gathered by the tribes.
  - Ranchers tend cattle on land that has been made available through large-scale destruction of trees.
  - Land has been made available by governments for small-scale farmers, many of whom have been encouraged to migrate from cities to **rural** areas.

Throughout equatorial areas of the world, during the last 50 years thousands of square kilometres of trees have been destroyed in various ways and for a variety of reasons.

## Methods of destruction

- Destruction by fire, which may be the easiest and quickest way to destroy forests.
- Trees being cut down by **logging** workers, since hardwoods such as teak and mahogany can be sold for large sums of money abroad.
- Trees drowning due to valleys being flooded to create reservoirs for multi-purpose water schemes.

Despite the damaging effects on the environment, people in countries such as Brazil and the Democratic Republic of the Congo and parts of India and south-east Asia continue to destroy the forest.

## Reasons for destruction

- Cutting down hardwood trees for timber for export.
- Clearing areas for small-scale farmers attracted to the rural areas from cities through government incentives.
- Clearing trees for mining a variety of minerals, such as iron ore, copper, bauxite and gold, using, for example, high-powered water jets.
- Using trees for fuel, for example charcoal for local industry such as iron smelting.
- Clearing large areas of forest to create grazing land for **cattle ranching**.
- Flooding valleys to create reservoirs for **hydroelectric power (HEP) schemes**.

All of these reasons are aimed at obtaining money from destructive activities.

## The impact of deforestation on people and the environment

- Destruction of trees can have a devastating effect on people.
- Thousands of different species of animals and insects are destroyed each year.

- Thousands of different species of plants are also destroyed each year.
- The Earth's atmosphere is seriously affected through the release of carbon dioxide and the loss of oxygen.
- **Deforestation** is known to contribute to **global warming** and the **greenhouse effect**, leading to increases in sea levels throughout the world.
- Rivers become polluted from mining enterprises, killing fish and affecting food supply and the welfare of tribes living in the forest.
- Without trees to interrupt run-off of rainwater, serious flooding occurs in other areas, causing widespread disaster and thousands of deaths.

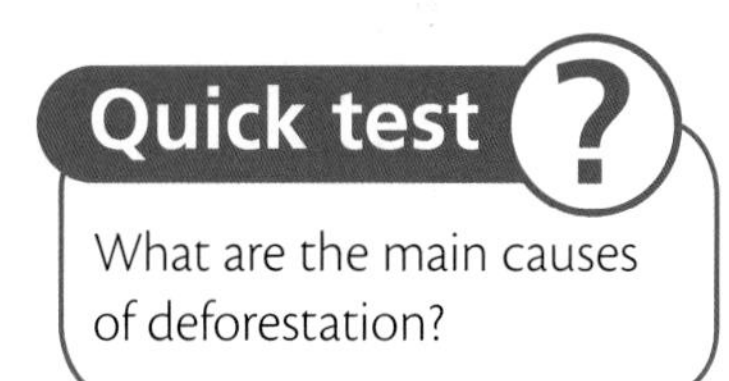


What are the main causes of deforestation?

## Management strategies to reduce deforestation

- Governments passing laws to protect forests by limiting the amount of land that can be used for activities such as mining and ranching. These laws are often very difficult to enforce.
- Worldwide campaigns by various protest groups, such as Greenpeace, to save the forests, for example, through protests to specific governments.
- Encouraging other commercial developments within the forests by buying forest products such as tropical fruits.
- There are many who live in the forest who can work in harmony with it, including tribes who hunt and gather and are subsistence farmers, rubber tappers, farmers, and those who replant areas that have been cut down with young trees.

Unfortunately, the process is continuing faster than the efforts to reduce deforestation.

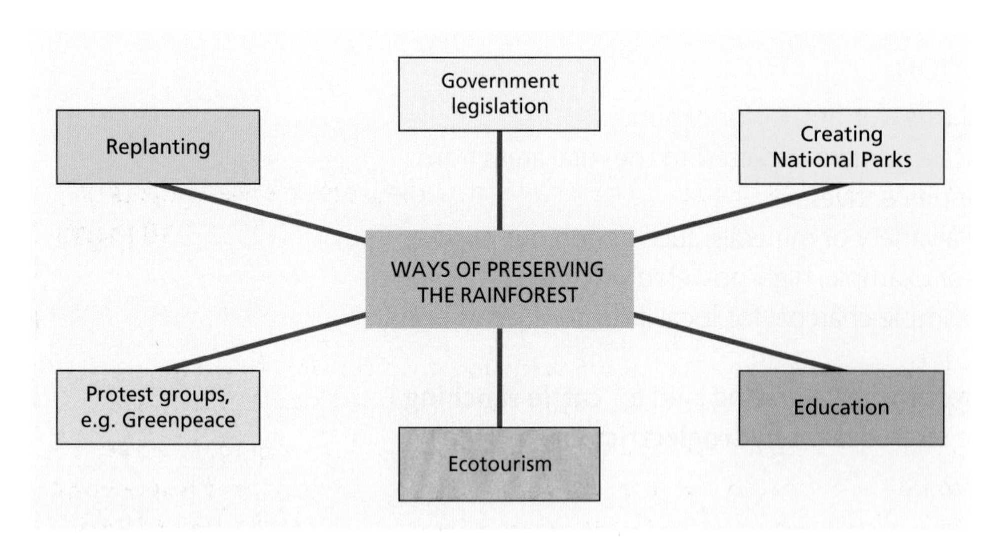


**Figure 2.5** Methods of preserving the rainforest

## Case study: deforestation in the Amazon Basin

For this area, the causes of deforestation include those that have already been discussed in this topic. This case study concentrates on the impact of this process and management strategies to deal with the issue.

**Impact on people, and social and economic consequences:**

- Deforestation has destroyed local people's way of life and caused clashes between locals and incomers.
- It has destroyed formerly sustainable activities, such as rubber-tapping.
- A reduction of fallow periods has reduced yields, which has led to food shortages.
- The interests of large businesses act in **conflict** with local interests.
- Local people have migrated away from traditional habitats, increasing poverty and social deprivation.

**Impact on land and environmental consequences:**

- Deforestation impacts on the nutrient cycle, which causes leaching and laterisation of soils exposed to the elements.
- There is increased run-off of groundwater and flooding.
- Flooding can remove soil.
- Some activities, such as mining, cause increased pollution.
- There is a loss of wildlife **habitats**.
- There are wider effects on global climates through the greenhouse effect.
- There may be an impact on local climate due to lack of moisture recycling.

**Strategies to manage and reduce deforestation in the Amazon Basin:**

- **Reforestation** with mixed trees.
- The use of crop rotation by farmers.
- The purchase of forest areas by **conservation** groups.
- Returning forests to native peoples.
- Several schemes are very effective, but outside interests in mining, ranching and so on often take precedence over conservation measures.
- There have been attempts to control this through government legislation, but the impact has been limited due to economic demands for development.

**Example**

Referring to **either** a named rainforest **or** a semi-arid area you have studied, **discuss** the impacts of land degradation on the area. **10 marks**

## Sample answer

*In the Amazon rainforest, trees have been cut down for wood supplies and to make way for mining and cattle ranching. This has meant that animals and plants have been destroyed as have the habitats of many species of animal. (✓) Local tribespeople have lost their homes and have been forced to move elsewhere. (✓) The loss of trees has meant that there are no trees to protect the soil from erosion from heavy rainfall (✓) and there are no roots to bind the soil together. (✓) Soil is washed away causing erosion (✓) and is deposited in rivers, polluting local rivers. (✓) This kills fish and local people lose a source of food supply. (✓) Destroying the trees affects the atmosphere since trees give off oxygen to the atmosphere. Burning trees adds carbon dioxide to the air causing pollution which can lead to global warming. (✓)*

⇨

## Comment and marks

This answer contains many good points about the impact of rainforest destruction. The answer refers to a case study – the Amazon rainforest – and therefore is able to gain full marks. The reference to destruction of plants/animals and habitats merits marks. A further mark is obtained for reference to local people being forced to move. Additional marks are gained for the references to soil erosion, pollution of rivers and the loss of food supply. The final mark is given for the mention of loss of oxygen and global warming.

The answer has covered impact on both people and landscape and obtains **8 marks out of the 10** available.

# Degradation of semi-arid areas

## Desertification background details

This background information will help your understanding of management strategies.

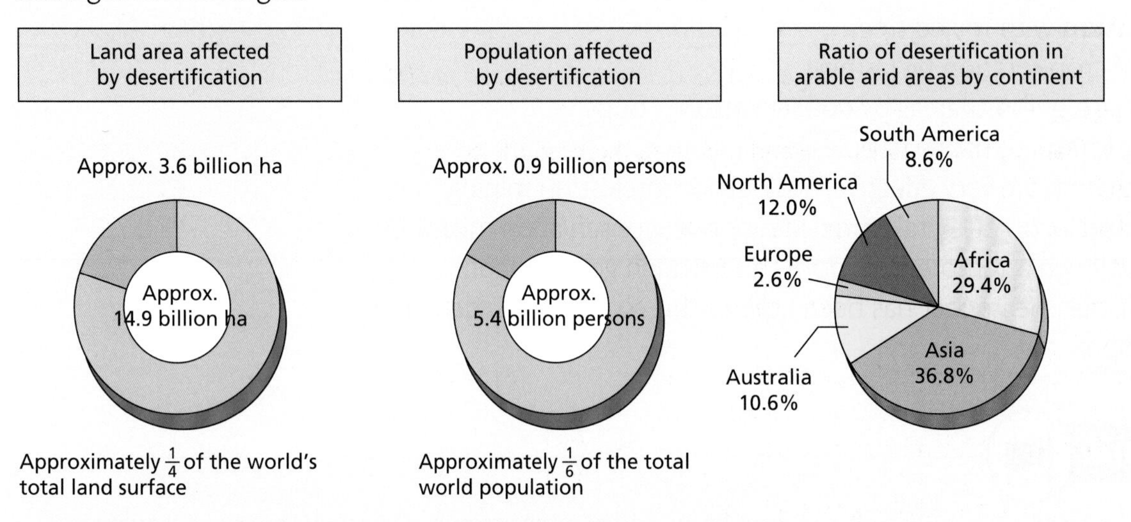


Source: The Ministry of the Environment from the Desertification Control Bulletin

**Figure 2.6** The impact of desertification

### Physical features

- Not all desert areas consist of sand. Many consist of dry rocks and tracts of sand.
- The climate consists of high temperatures throughout the year; in some cases, in excess of 40°C, and low rainfall.
- Soils are known as desert soils, which, due to the lack of moisture, are generally infertile.
- Oases provide the main natural water source in deserts.
- Whenever rainfall does occur it often results in flash floods, which wash away topsoil, leaving the land even more infertile.
- Areas within deserts are characterised by dried-up river beds (**wadis**).

*Key point*

For a semi-arid area you have studied you may be asked to explain the impact of rural land degradation and the management strategies employed.

## Human features

- Population density is generally low, apart from the **settlements** that are based around oases.
- The main type of farming is pastoral **nomadism**. This involves tribesmen who travel with their livestock throughout the year in search of new pastures for their animals, which consist mainly of sheep and goats.
- Where water is made available through **irrigation**, crop farming is possible, and crops include wheat, maize, cotton and tobacco.
- This farming is usually found in the floodplains of rivers such as the Nile.

## The process of desertification

- **Desertification** is the process by which formerly productive land has been turned into desert.
- At the edges of hot deserts throughout the world the deserts are spreading into areas that were formerly settled and farmed and provided a living for many people.
- Although there are physical factors responsible for this process, including prolonged periods of **drought**, human factors are probably most responsible for this process occurring.

## The causes of desertification

- The main physical cause of desertification is prolonged absence of rainfall; in other words, prolonged drought.
- As a result, plants and vegetation die and therefore there is little cover over soils.
- A second factor is the effect of wind, which blows away dead vegetation and topsoil, eroding the land and depositing fine sand elsewhere.
- Insects may contribute by eating vegetation, which helps to remove soil protection.
- Human activities add to all these physical processes, which help to increase the process. These activities include, for example:
  - deforestation, by which trees may be cut down for firewood
  - farmers allowing their animals, such as sheep and goats, to overgraze the vegetation and lead to more **soil erosion**
  - farmers **overcultivating** the land, removing any goodness in the soil and preventing the soil from supporting any more crops.
- People do this because there is a desperate need for food during periods of extensive drought and through a lack of expertise in proper farming techniques.

## The impact of desertification

- Without anyone to care for the land, land that was formerly fertile and settled gradually turns to desert, thus increasing the extent of deserts.
- Farmers are unable to graze animals.
- Soils that are exposed to the elements are eroded and therefore unable to sustain crop growth.
- Since the land cannot sustain any further farming, local people have little option but to move away to areas where the soil might be more fertile.
- Villages are abandoned.
- People may migrate to other areas.
- Traditional farming, for example pastoral nomadism, may disappear.

# Management strategies to reduce desertification and their effectiveness

- The most effective measure that can be taken to reduce this process is to bring water to dry areas. This can be done through various forms of irrigation, especially digging wells and ditches, or through large-scale multi-purpose water projects, to bring much-needed water to dried-up land.
- Governments can help farmers by providing people to advise them on better farming methods, such as using fertilisers and 'miracle seeds' where possible.
- Fencing off grassland areas avoids **overgrazing**.
- Using **soil conservation** methods.
- Planting young trees to act as **windbreaks**.
- Avoiding any of the actions that lead to speeding up of the process.
- Much-needed financial assistance provided to governments through international aid agencies such as the World Health Organization, United Nations Food and Agricultural Organization and the World Bank and loans and aid programmes from more developed countries throughout the world.
- In the Sahel, magic stones (diguettes) are used. These are lines of stones placed along the contours of the land that trap run-off from rain water as well as soil. This is particularly useful following seasonal rainfall in the Sahel.
- Another method used is animal fences. Animals such as goats will eat anything and graze the vegetation down to the roots (overgrazing). This prevents regrowth, and soil without roots to bind it together gets blown away. Animal fences prevent this by keeping animals out of fragile areas, giving the land time to recover and regrow, or grazing can be controlled to prevent overgrazing. Rotating the areas fenced off from animals allows herds to be sustained without long-term damage to the soil.
- Reducing the size of herds, focusing on quality rather than quantity, decreases the grazing pressure on the land. With fewer hoofs trampling the soil, it becomes less compacted and allows water to infiltrate the soil.
- Farmers should be educated so they are aware of the causes and consequences of land **degradation** and of better methods of farming, for example drip irrigation.
- Some of these methods have been successful. Stone lines have little cost. Farmers can help each other. In Mali and Burkina Faso, for example, this method has been successful in increasing some crop yields by as much as 50 per cent.
- Managed grazing areas are successful if fencing is available and affordable.
- In some areas, the number of cattle owned is seen as a status symbol, so the method of reducing herd size is not always successful. In addition, agreement needs to be reached with herders, which is not always possible, for example in the case of the settlement of Korr in Northern Kenya.
- Even if education is available, farmers very often cannot afford the techniques that are suggested, such as drip irrigation and fertilisers.

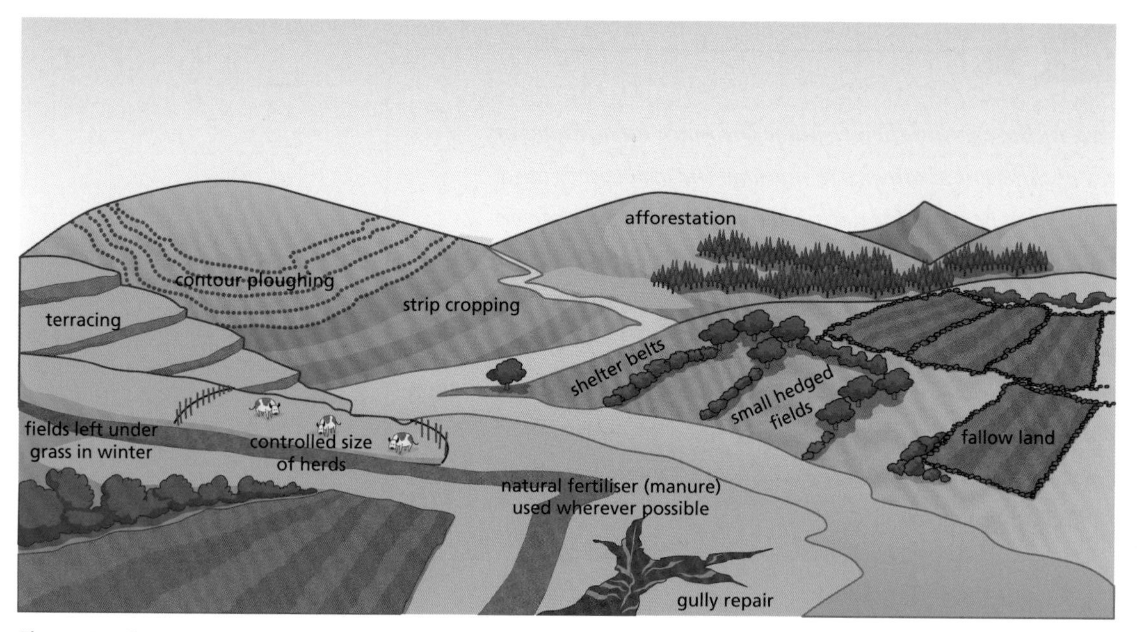


**Figure 2.7** Conservation measures to prevent desertification

# Case study: desertification in Africa north of the equator

## The impact on people, and social and economic consequences

- Crop failures, and consequently malnutrition, lead to major famines, for example in Ethiopia and Sudan. This leads to mass migrations, often to refugee camps or shanty towns on the edge of cities.
- There is often a collapse of traditional activities, for example nomadism, due to overgrazing and lack of water.
- There is increased pressure on land due to nomads settling in villages, which results in increased tensions between nomads and traditional farmers.
- There may be widespread desperate poverty and deprivation, which results in increased mortality rates, especially infant mortality rates.

## The impact on land, and environmental consequences

- There is a breakdown of soil structure, which causes an advance of the Sahara desert, resulting in the process of desertification.
- There is wind erosion of dried-out soils.
- There is also erosion from rains when they eventually arrive.
- Water tables are lowered.

## Management strategies for Africa north of the equator

- Building dams, for example in parts of Kenya.
- Planting trees as windbreaks.
- Stabilising dunes with grass, for example in parts of Mali.
- Terracing slopes to prevent erosion.
- Improved irrigation: Even small-scale schemes are very effective in increasing soil depth and crop yields.
- Controlled grazing and fencing, which are also very effective in preventing top vegetation being removed and exposing soil to wind erosion.

### Hints & tips

- *If you are asked in the examination to judge the main advantages or disadvantages of different strategies to manage the problem of land degradation, you may be provided with a resource such as a map or diagram.*
- *You must be able to use the information given in the source to support points that you make in your answer.*
- *Marks are gained for selecting the appropriate data in support of your opinions or choice.*
- *Use the number of marks allotted to the question as a guide to the depth and amount of detail required.*

# Rural issues

## Glaciated upland

### Case study 1: the Lake District

Although details on **land use** are not assessable, you need to be familiar with them to fully answer any questions relating to rural **land-use conflicts** and management strategies related to glaciated and coastal landscapes.

#### Key point

Referring to either a glaciated upland area or a coastal area you have studied in the UK, you should be able to discuss rural land-use conflicts within these areas and strategies employed to manage these landscapes.

**Background:**

- Three main rock types make up this area; namely, igneous rocks such as granites; metamorphic rocks such as slate; and sedimentary rocks such as grits and limestone.
- This area contains England's best example of an upland glaciated area with a wide variety of features including pyramidal peaks, tarns, ribbon-shaped lakes, U-shaped valleys and hanging valleys.
- The most famous lakes include Windermere, Ullswater, Coniston Water and Thirlmere.
- The Lake District is one of the UK's most popular **National Parks**. It has spectacular examples of glaciated scenery.
- The upland glaciated areas such as Helvellyn, Scafell Pike and Striding Edge are popular with hill walkers and tourists, who climb the peaks to admire the views and take photographs of the surrounding area.
- The many lakes, including Windermere and Bassenthwaite, provide opportunities for water-related activities.
- The glaciated valleys, such as Langdale, are popular with ramblers and picnickers.
- There are many small picturesque villages, such as Ambleside. This is popular with visitors because the buildings are old and built from local stone and there are tea shops, cafés and gift shops.
- Areas of forestry provide attractive natural environments and mountain-bike trails and opportunities to study wildlife in their natural habitat.
- The area also contains places of historical interest, such as William Wordsworth's house in Cockermouth and Hill Top Farm in Sawrey, where Beatrix Potter lived.

- Efforts are made to protect the physical environment by **National Park Authorities** and other bodies.

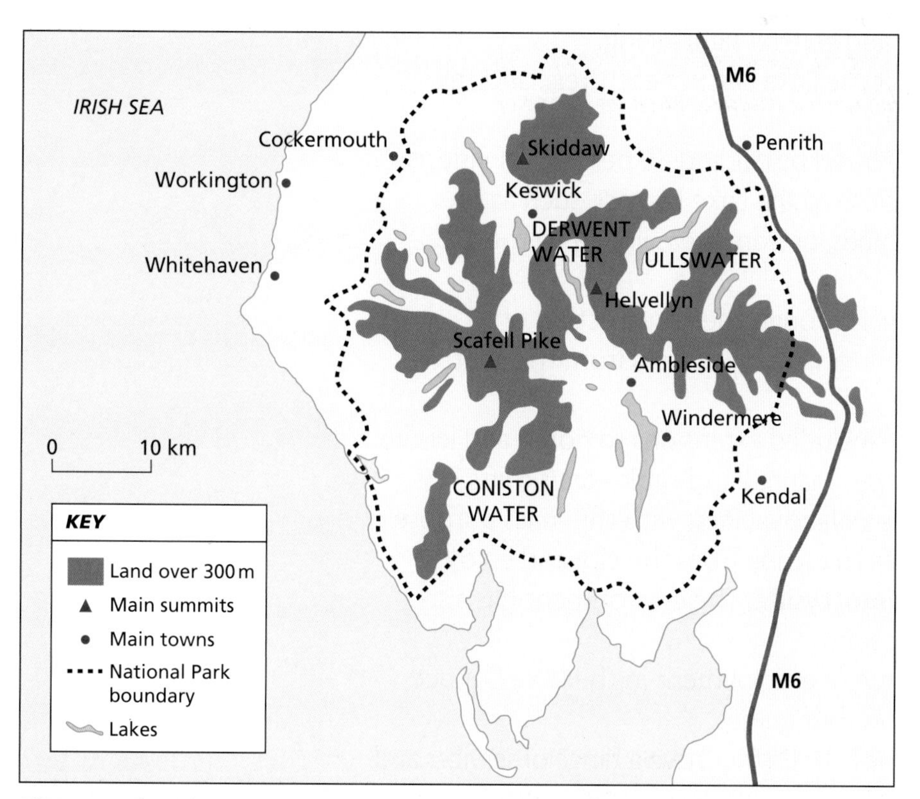


**Figure 2.8** The Lake District

**The following land uses are found in this area:**

- **Farming:** Due to the steepness of the slopes, the cold temperatures and high rainfall, which limit the growing season and affect soil fertility, the area is unsuited to crop farming, since it would be almost impossible to use machinery such as combine harvesters.

  The only type of farming that is feasible is hill sheep farming, with cattle occasionally being raised on lower, less steep land.
- **Forestry:** Large plantations of coniferous forests are found in this area. This economic activity is suited to this area of steep slopes, poor soils and inhospitable climate and provides employment in the timber, paper and furniture industries. The trees also protect the slopes from soil erosion.
- **Industry:** Due to the lack of flat land, not much manufacturing has been attracted to the area. The main type of industry is quarrying: of granite, slate for roads and roofs and limestone for use in steel-making elsewhere. The number of operating quarries has gradually and seriously declined in recent years.
- **Water supply:** The lakes have supplied Manchester with water for over 100 years despite being over 150 kilometres distant from the city. These lakes are natural reservoirs in an area of high rainfall and are much more economical for Manchester to use than building man-made reservoirs. The lakes supply up to 30 per cent of the water needs of this part of Britain.

- **Recreation and leisure/tourism:** The area is very attractive to tourists as it offers a variety of physical attractions, such as the mountains and lakes for activities including hill walking, mountain climbing, adventure holidays, watersports, fishing and general sightseeing.

  Over 12 million people visit the Lake District each year, and the number is increasing annually.

  Most visitors (90 per cent) travel by car and 50 per cent of visitors are either from regions fairly close to the Lake District, such as Newcastle, Manchester and Leeds, or from areas linked by the M6, such as Birmingham.

  Access has been made easier through the construction of motorways such as the M6, bringing many visitors from the south of Britain to the area.

  Recent developments have included extensions to hotels and leisure complexes, timeshare complexes, marinas and cable cars/ski lifts.

  The area was designated as a National Park with the main purpose of offering city dwellers a place to escape from the city and enjoy the benefits of a protected **countryside**. The environment has also benefited from National Park status.

  Tourism is the biggest source of employment in the Lake District. The park employs wardens and tour guides.

  The towns and villages of the Lake District have a range of services and facilities that provide employment for locals on a full- and part-time basis:
  - Local crafts can be sold in souvenir shops.
  - Farmers can provide the cafés and restaurants with fresh produce.
  - Tourists spend money in the area and some of this can then be used to improve roads and services for locals and visitors. This helps to stop rural depopulation, whereas other parts of Britain with similar physical characteristics, such as north-west Scotland, have become seriously depopulated.
  - Many jobs have been provided in the area by tourism, and in some parts 50 per cent of the workforce is employed in the tourist industry.
  - The tourist industry has helped local employment opportunities by making more jobs available. This is especially important since tourist jobs have replaced jobs lost in agriculture.
  - There are also many other businesses that rely partly on tourism for employment, such as the building and catering industries.
- **Housing:** Housing is in short supply in the area and house prices have increased dramatically.

  Much of the existing housing is used for 'second homes', which means that local people, especially young people, have to move out. This causes resentment among the local population due to the loss of trade to local businesses when second homes are left empty during the year.

In commercial terms, these land uses are very important to the local economy. They provide employment and income in a variety of ways. However, these land uses have had a significant impact on the natural environment, causing increases in pollution, traffic congestion, footpath erosion and changes to the traditional rural character of many villages.

**Land-use conflicts within the Lake District include conflict between tourism and farming due to factors such as:**

- Increased traffic congestion, due to tractors holding up traffic and heavy use of small rural roads by tourist traffic.
- Damage caused by tourists in farming areas by litter, damage to fields, increased pollution, gates left open and animals worried by family pets.
- Farmers have blocked access to public footpaths.

**Other land-use conflicts include:**

- Conflict between **farming** and other major land users, especially **water boards** who flood areas of valuable farmland and **developers** looking for land for industrial purposes.
- The increased use of lakes for water-skiing, power-boating and jet-skiing creates conflict between these activities and other lake users, such as swimmers and anglers, and protection of the natural environment.

**Management strategies that have been and are employed to resolve land-use conflicts in the Lake District include:**

- Legislation such as the Green Belt Act enforced both to protect and to conserve the area from industrial and urban developments.
- Some strategies involving partnership between the National Park Authorities, the National Trust and the tourist and hotel industries to encourage sustainable tourism.
  - This involves raising visitor awareness, raising money to restore and conserve the landscape and ensuring that tourism and conservation work together to benefit the local community.
- Strict planning laws observed to ensure that any development is both in character with the area and environmentally sustainable.
  - This means that new developments should not adversely affect the environment, economy or social character of the Lake District in the long term.

**Efforts by public and voluntary bodies to help resolve issues in the Lake District involve:**

- The National Park Authority adopting many measures to protect and conserve the natural environment, including the landscape, local vegetation and wildlife.
  - These measures include those noted earlier, such as zoning of activities, reducing traffic congestion, ensuring a reasonable amount of new housing is sold only to local people and providing information and education centres for visitors.
- The National Trust owning and protecting approximately 25 per cent of the land in the Lake District.
  - The Trust provides footpaths for tourists, thus reducing the risk of erosion. It also maintains dry stone dykes and the habitats of wildlife in the area, and therefore helps to reduce conflict between tourists and local farmers

**Hints & tips**

*When answering a question on this topic, make sure that you mention a particular upland area and identify specific places within it, such as towns and villages, and local landscape features such as lakes or mountains.*

# Coastal landscapes

## Case study 2: the Dorset coast

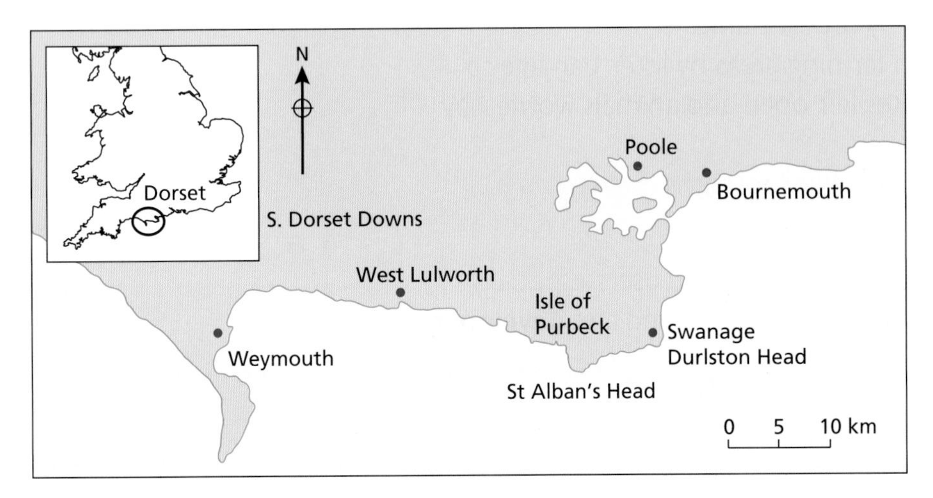


**Figure 2.9** The Dorset coastal area

> **Key point**
>
> Referring to a coastal area you have studied, you should be able to explain the main land-use conflicts and how they are managed.

All of the activities shown in Figure 2.10 are to some extent in conflict with the natural environment and threaten to change the natural balance and ecological diversity of the area.

In addition, the coastline is continually under threat from natural forces such as waves, currents and groundwater movement.

The coastline needs to be managed to sustain human activities threatened by marine erosion, to preserve coasts for conservation reasons and to preserve them from development.

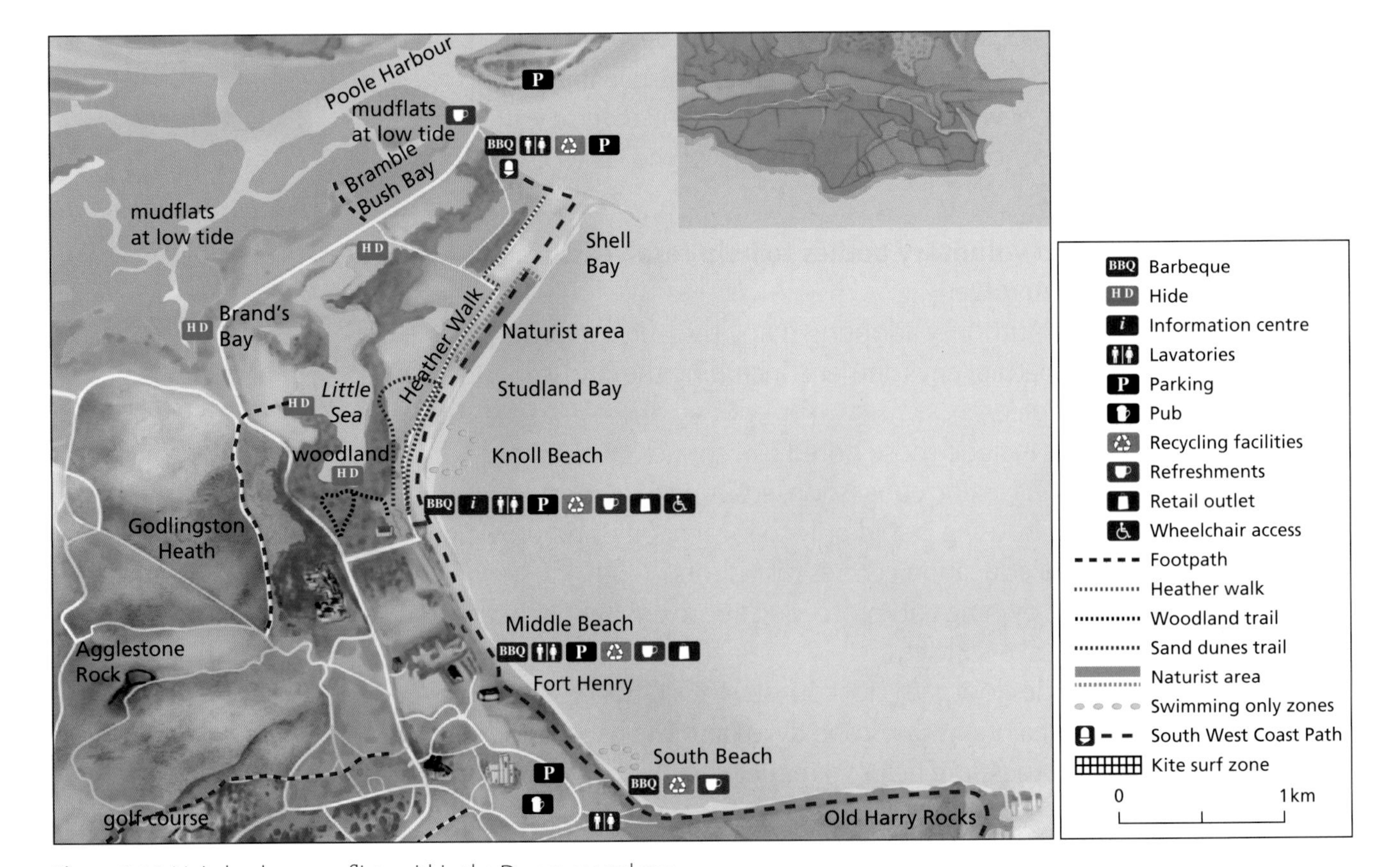


**Figure 2.10** Main land-use conflicts within the Dorset coastal area

**Management strategies employed to resolve these conflicts:**
A document outlining the strategy for sustaining and improving the quality of the environment, called the 'New Forest 2000', was published in 1990.

- Measures employed in this strategy included efforts to reduce pollution levels, protect the scenic beauty, improve the appearance of the coast, maintain the economy, protect the coastline, educate the public and conserve features of historic and archaeological interest.
- There have also been attempts to protect coastlines from flooding using dykes and flooding walls.
- Responsibility for managing coasts generally lies with three agencies; namely, the Environment Agency, MAFF (national government agency for coasts) and district councils.
- Coastal defence strategies include measures such as sea walls, using large irregular rocks, gabions (wire baskets filled with rubble), groynes and embankments. Each of these measures has several advantages and disadvantages, not the least of which is cost. Some are relatively cheap while others, such as sea walls, can be very expensive to build and maintain.
- Nature reserves have been created to protect wildlife, and two **country parks** have been created to encourage sustainable tourism.
- Public authorities, which operate the nature reserves and the country parks, manage these strategies, and the area has achieved the status – if not quite the title – of a National Park.
- These efforts in the New Forest area have met with considerable success.

**The role public and voluntary bodies in the Dorset coastal area have played in dealing with land-use conflicts:**

- Public and voluntary bodies, other than those associated with nature reserves, involved in efforts to resolve land-use issues include various Coastal Management Authorities:
  - The designation of Heritage Coast Status for the area.
  - The work of the Environment Agency, National Trust and National Nature Reserves.
  - The designation as a Site of Special Scientific Interest (SSSI) and as a Special Marine Conservation Area.

  All of these bodies make great efforts to ensure that the natural environment of this area is protected as much as possible from the pressures and threats outlined in Figure 2.10.

**Social, economic and environmental impacts the main land uses have had on the area:**

- The social impact is felt in terms of the area being a magnet for population since it is very attractive to tourists and as a place to live, work and retire.
- Some 17 million tourists are attracted to the area each year.
- There are over 200,000 educational visitors catered for every year.
- The area benefits economically from the wide range of activities practised there, including tourism, industry, forestry and farming.

These activities generate a great deal of money for the people who live and work in the area.

- Environmental effects of the demands from tourism, port and ferry services and increased traffic congestion include footpath erosion; demand for land for car parks and other amenities; threats to wildlife habitats; danger of marine pollution; conflicts between fishing, water sports and marine archaeology; and the presence of the UK's longest onshore oilfield in the area.

*When answering a question on this topic, make sure that you refer to a specific coastal area and identify specific places within it.*

## Key words and associated terms

**Cattle ranching:** Rearing of large herds of cattle on areas of cleared forests to provide beef for export.
**Deforestation:** Removal of trees, usually on a large scale.
**Degradation:** The process of reducing land that was formerly productive into unproductive land.
**Desertification:** The process of turning land that was formerly productive into desert.
**Drought:** Prolonged periods without rainfall, which may last from several months to several years.
**Global warming:** The heating up of the Earth's atmosphere by the Sun's rays due to the effect of atmospheric pollution.
**Greenhouse effect:** A rise in temperature in the Earth's atmosphere due to the effect of increased carbon dioxide and other gases in the atmosphere from various sources, such as industrial pollution, burning forests, car exhaust fumes and smoke from domestic chimneys.
**Irrigation:** An artificial way of providing water for farming from sources such as rivers, wells, canals and field sprays.
**Logging:** A commercial business that cuts down trees to provide timber for sale for different purposes.
**Nomadism:** The system whereby people migrate with animals throughout the year to find new areas for grazing livestock; also known as 'pastoral nomadism'.
**Overcultivating:** Growing crops continually to the extent that the soil becomes exhausted of nutrients, infertile and unable to sustain further growth.
**Overgrazing:** Allowing animals to overeat grass to the extent that the underlying soil is exposed and cannot sustain further growth.
**Reforestation:** The process of replanting trees in former forested areas.
**Shelter belt:** A line of trees planted to provide shelter from the wind for fields by interrupting the flow of wind.
**Soil conservation:** Attempts to protect soil from damage using methods such as fertilisers, irrigation, shelter belts and ploughing along contours and using terraces to conserve soil.
**Soil erosion:** The process by which the topsoil is removed, leaving the land infertile.
**Windbreaks:** Trees that interrupt the flow of wind so as to protect fields and their crops.

### Land-use conflicts

**Conflict:** Happens when two or more land users disagree as to the best use of the land. Often one land use is totally at odds with another, such as industrial activities perhaps spoiling and polluting areas used for farming.
**Conservation:** Efforts to maintain the basic beauty and attractiveness of areas, both in the countryside and towns.
**Country park:** An area in the countryside surrounding a town or city that has been set aside for people to visit as a park.
**Countryside:** The majority of land area that has not been used for towns or cities.
**Habitats:** Places where animals and birds live, for example hedgerows, fields and woodland. Often these are lost when land is developed.
**Hydroelectric power (HEP) schemes:** Hydroelectric power schemes built for the purpose of creating electricity using water in reservoirs as a source of power to drive turbines. The electricity is fed into the national electricity grid.

**Land use:** How humans make use of the physical landscape, for example forestry, farming, industry and settlement.

**Land-use conflict:** Occurs when different activities compete with each other to make use of the land, for example farming and tourism.

**National Park Authority:** An organisation that looks after areas that have been set aside throughout Britain for public recreation and enjoyment. Its aims also include protecting areas of outstanding scenic beauty.

**National Parks:** Places of outstanding scenery set aside and protected for the purpose of attracting people from urban areas for leisure and recreational purposes. These areas are looked after by the government agency, the National Park Authority (see above). There are thirteen National Parks in England and Wales and two in Scotland.

**Rural:** Another name for countryside areas.

**Settlement:** Places where people live and work, ranging from small villages to large towns and cities.

# Chapter 2.3
# Urban

## Key points 

For this chapter you must know about the need for management of recent urban change in a developed- and developing-world city. The two areas of change that you should know about are **housing** and **transport**.
You also need to know about the management strategies employed and the impact of these management strategies.

# Change in developed-world cities

## The need for managed change and its impact

### Change in housing areas

- Different housing areas can be identified by their street patterns.
- Low-cost housing areas have houses that were built in a gridiron pattern, often with small narrow streets. They are situated very close to the older industrial areas, to allow workers quick and easy access to their place of employment.
- The environments of older low-cost housing areas are not very pleasant.
- The housing almost certainly consists of tenements, which are high-density housing.
- Many of these tenements show signs of age and the effects of years of pollution from nearby factories.
- Some cities have demolished many of the older tenements and have replaced them with newer flats; very often high-rise flats.
- Many people have been rehoused in schemes built on the boundaries of cities. These are recognisable by the layout, which is often very similar to the high-density older tenement areas of the **inner city**.

#### Key point 

To understand why there was a need for managed change in developed-world cities, it is important to appreciate where and why this change occurred.

## Housing types

In developed-world cities, four main types of housing areas have been identified:

1. areas of low-cost housing
2. areas of medium-cost housing
3. areas of high-cost housing
4. areas of housing on the **rural/urban fringe** (outskirts) of cities.

Most of the changes that have taken place in developed-world cities occurred in low-cost housing and urban/rural fringe housing areas.

#### Key point 

When studying this topic you should concentrate on these particular housing areas: low-cost housing and urban/rural fringe housing areas.

# Change in low-cost housing areas of developed-world cities

This zone contains mainly less expensive, often older, houses. Since the zone is closest to the industrial zone, the houses would originally have been used primarily to house the workers of the industries in industrial areas.

Much of the change needed was due to the demands to improve poor, low-cost housing areas in the inner city.

- Many parts of cities have buildings that, through age and lack of care, have fallen into disrepair.
- These buildings may be industrial buildings, housing, office blocks or even shopping areas. Their visual appearance may be unsightly and, more importantly, they may be unsafe for use or habitation.
- If they are no longer in use they may become **decayed**. Often these buildings are left for many years before action is taken.
- Common sites for them include near railway lines, canals, along the banks of rivers and perhaps at the edges of the central business area.
- The problem was often a lack of money for demolition and redevelopment.

## Management strategies adopted to secure change in the inner-city areas

- Many city authorities and councils have made great efforts to remove these older dilapidated buildings and replace them with newer buildings.
- New buildings include office blocks, new housing, recreational centres and occasionally under-cover shopping centres.
- These changes have often been achieved through the financial assistance of government grants and even European Union grants.

## The impact of change in the inner-city housing areas

- Regeneration schemes in some inner-city housing areas radically changed the quality of housing and the local environment of many housing zones in cities.
- Older properties, such as nineteenth-century tenements, were demolished and replaced with new housing.
- This included demolishing high-rise flats built during the 1960s, which caused great problems for residents, particularly young families and elderly people.
- Many families and older communities were broken up and redistributed to areas throughout the city and consequently lost a sense of community spirit and security.

# Medium- and high-cost housing areas

There was much less need for change in the medium- and high-cost housing areas of cities. These areas did not experience the problems that occurred in the inner-city and rural/urban fringe areas.

# Managed change in rural/urban fringe areas

Change was required due to the following factors:

- The population of many cities had decreased as many people left to live in various settlements outside the main city.
- Large numbers of people were prepared to pay the cost of increased travel in terms of time and money to live in what they considered a better environment. Small rural villages were essentially colonised and have become commuter settlements.
- The populations of these former rural villages have often grown enormously. The services provided for their populations often bear no relation to their population size. Residents commute to work and purchase most of their needs outside the villages.
- In other cases, large numbers of people from cities in Britain were offered the chance to move to new towns, with the promise of employment as well as housing.
- Many new towns were built throughout Britain and included a wide variety of services, entertainment, schools and, most important, a wide selection of houses.
- These new towns were very popular, for example East Kilbride is the largest and most successful of Scotland's new towns and has a population of about 75,000.
- The result of these movements was that the population of cities and the income from this population decreased alarmingly.
- Cities need people to support their services.

## Strategies used to manage change in outer fringe areas

- Many cities have tried to attract people back by investing in new and better housing and shopping areas and by trying to attract new industry. This is an ongoing struggle but a number of large cities, including Birmingham, Glasgow and Edinburgh, have been successful.
- A great deal of money has been invested in upgrading older properties close to the centres of many cities and making them attractive housing areas.
- The idea was to attract more and more people from the commuter settlements into the city by proving that these people can benefit from living in a reasonably pleasant environment without having to pay large sums of money on transport costs and having to endure long and difficult journeys to work each day.
- Loss of population due to the development of new towns and commuter belts has created financial problems for city councils, traffic problems and loss of business for city-centre shopping areas.

- On the rural/urban fringe, developments such as housing and new industrial estates have led to urban sprawl and the loss of rural land, which has caused conflict between traditional rural activities, such as farming, and new developments.
- Green-belt legislation introduced in the 1950s was an attempt to curb development in rural fringe areas and has been quite successful.
- Limitations on the number and type of buildings have kept the rural environment protected from the worst excesses of city developers.

## The impact of change in rural/urban fringe areas

- Property speculation and compulsory purchase of land by developers often led to a decline in the quality of the rural/urban fringe around many cities, which resulted in loss of farmland and recreational land.
- As the commuter belt expanded there was ever-increasing demand for new houses. This in turn led to increase in traffic on rural roads throughout the week from Monday to Friday.
- Green-belt strategies involved planning restrictions and restrictions on developments such as housing, industry, landfill sites and recreational centres.
- Smaller towns and villages were identified for growth.
- In central Scotland, new towns and overspill areas such as East Kilbride were used as growth centres to limit further development within the rural/urban fringe.
- However, small rural villages remained popular with people wishing to leave the city and were still a target for building developers.
- The needs of industry and opportunity for offering employment had to be balanced with the desire to protect the rural/urban fringe.
- Protection efforts, despite green-belt legislation, have not always been successful in limiting the negative impact of some developments.

# Case study 1: Glasgow

## The need for change to housing in Glasgow

Change to housing in Glasgow was needed for several reasons:

- Many parts of Glasgow had buildings that, through age and lack of care, had fallen into disrepair.
- Many houses were contained tenements built in the nineteenth century to house immigrant workers, particularly from Ireland.
- These tenements were cheap to build and many were crammed into large blocks in a gridiron pattern.
- Many tenements had no toilet facilities and tenants had to share a common toilet with others.
- Streets were narrow and highly congested.
- The houses were built next to docks and heavy industry for the convenience of the workers.
- The housing areas were often polluted by river pollution from industries discharging waste, urban blight caused by **derelict** buildings, litter, vandalism and air pollution from industries and traffic.

For a named city in the developed world, you should be able to explain the need for managed change to areas of housing.

- Pollution from industries, motor vehicles or neglect accumulated to such an extent that it not only destroyed the appearance of some parts of the city, but the pollutants also became a health hazard.
- Many of the large industrial areas of the city were abandoned.
- Factories were demolished for safety reasons, but the sites were left as 'gap sites'.
- Glasgow, which once had port and shipbuilding industries, was left with large areas of dockland that were no longer in use.
- Like many cities in this situation, abandoned buildings included industrial buildings, housing, office blocks and even shopping areas.
- Their visual appearance became unsightly and, more importantly, they were unsafe for use or habitation.
- Those that were no longer in use decayed, and often these buildings were left for many years before action was taken.
- Common sites for them included near railway lines, canals, inner-city housing areas, along the banks of rivers and at the edges of the central business area.
- The problem was a lack of money for demolition and redevelopment.
- In the years since the end of the Second World War, many of Britain's older, traditional industries have gone into economic decline, including shipbuilding, iron and steel, textiles and heavy engineering including those in Glasgow.

## Strategies involved in changing housing in Glasgow

- Like many cities, Glasgow city authorities made great efforts to remove abandoned and decayed buildings and replace them with newer buildings.
- New buildings included new housing areas such as New Gorbals, where old tenements were demolished and replaced by high-rise flats.
- Changes have often been achieved through the financial assistance of government grants and even European Union grants.
- New housing schemes were built on the outskirts of Glasgow, including Castlemilk, Easterhouse and Drumchapel, to rehouse families who lived in the old tenements. New towns such as East Kilbride and Cumbernauld were also built to house the overspill. These council-house schemes met with mixed success and impact on the inhabitants.
- Former dockland areas were developed for private housing and entertainment, such as the Quay complex in the south of Glasgow.
- In some areas like the Merchant City, new private flats have been built.
- The original policy involved the demolition of older houses, but this is now regarded as short-sighted and wasteful, and since the 1980s the policy has been one of urban **renewal and regeneration**.
- Many of Glasgow's old tenements have been refurbished into desirable accommodation, rather than demolished. They have been given new roofs and windows and sandblasted on the outside, while receiving new kitchens and bathrooms inside.

- Problems due to pollution have been successfully tackled in Glasgow due to measures such as clean-up programmes on the Rivers Clyde and Forth and the Clyde canal project, and **ring roads** and pedestrianisation of areas within the CBD, which divert traffic from the city centre.
- The City Council employs environmental health departments to implement laws and regulations to monitor and control pollution. Although this is costly, the benefits are obvious to the residents of cities.
- Glasgow improved facilities to such an extent that it has won awards such as the British city of architecture, culture awards and the awarding of the city as the host for the 2014 Commonwealth Games.

## The impact of the management strategies

- New council schemes like Castlemilk were partially successful. They provided people with a newer home with more modern facilities. The houses were more spacious and there were gardens for children to play.
- However, many older residents of the inner city were reluctant to move away from their friends and the amenities of the city and families were split up.
- Castlemilk was built at some distance from the city centre and so it costs more money and time to travel in and out of the city.
- Very few amenities were included in the construction of the new council estates.
- High-rise flats were built to replace the original housing in the Gorbals. This was of poor quality and many flats suffered from dampness. There was no place for children to play and older people became isolated in their flats if the lifts broke down.
- Noise and vandalism were common. These flats were not a success and have since been demolished.
- However, new towns like East Kilbride have been a success as there were employment opportunities for the residents who were rehoused there and the housing was of a reasonable standard. The city centre is accessible by train and bus.
- The refurbished tenements also a success as the residents were able to stay in the area they knew, and received a far bigger, more modern house.
- After the 2014 Commonwealth Games, the city authorities planned for the athletes' village to be transformed into 1400 houses of various designs, including flats and terraced and detached housing, for new residents.
- These new homes, for rent or sale, were to replace older and worn-out housing and bring new communities into different parts of Glasgow, including the east end.
- The new residents would have access to new shops, schools, community centres and sports facilities.

# Case study 2: inner-city changes in Paris

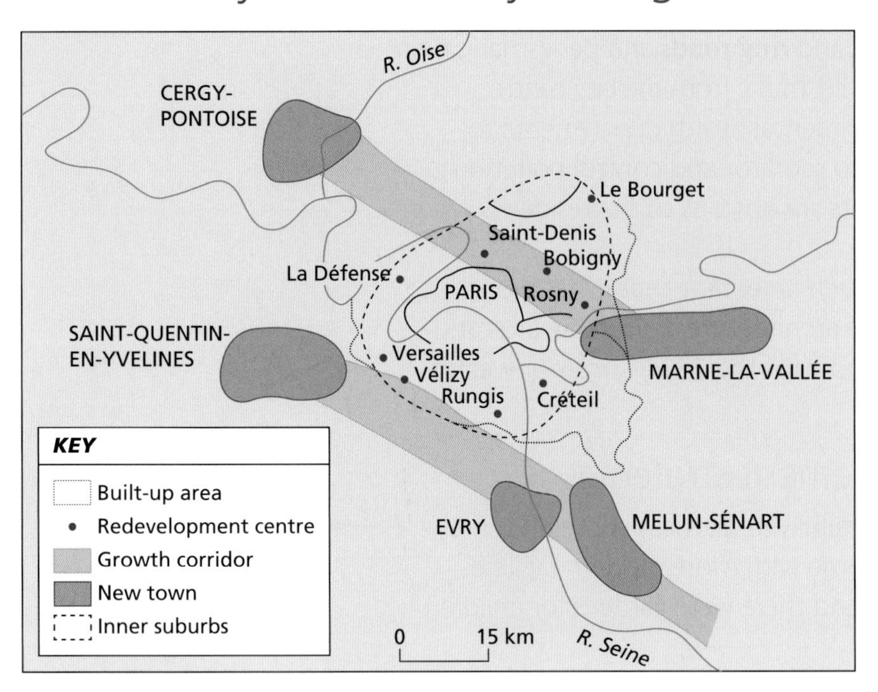


**Figure 2.11** Inner-city change in Paris

## Management strategies

- In Paris, several major changes have taken place within the inner city, most of these involved changes to housing.
- Old tenements, which often lacked basic facilities, were demolished and replaced with new and renovated housing.
- The local authorities in Paris set up regeneration schemes, such as La Défense, which involved pulling down delapidated tenements and replacing them with 25 tower blocks up to 40 storeys high.
- There are 35,000 people currently living in these high-rise blocks.
- La Défense provides new offices and shopping facilities in addition to the improved housing.
- The area is a traffic-free zone, with roads and car parks located underground, and the area is connected to the city centre by the underground Métro rail service.

## Gentrification of the inner-city area of Paris

- In Paris, the worst slum areas were pulled down or renovated and replaced by improved housing.
- Many wealthy middle-class people working in central Paris bought some of this housing cheaply.
- Houses were improved by the new owners, and this encouraged others to improve their properties.
- Gradually some parts of the inner city in Paris became fashionable to live in for the wealthy, including areas such as Belleville and Bercy.
- Advantages of living in these areas include easy and cheap access to work, shopping and entertainment in the central area.
- Commuting costs were reduced, making it easier to afford gentrified property in the inner city.

## Ghetto areas in Paris

- As redevelopment of the inner city took place in Paris, other areas, such as Sarcelles, housed considerable numbers of people in large housing schemes.
- These schemes became unpopular due to the lack of jobs, shops and public transport.
- As people moved away from the area they were replaced by immigrant families, especially from North Africa.
- Due to racism from landlords, many immigrants found it difficult to find accommodation in any areas other than the poorest.
- Since many immigrants preferred to live within their own ethnic community, with people who share the same language, culture and customs, the percentage of the immigrant community increased greatly.
- Some of the housing schemes became ghetto areas.
- Some immigrants, because of the lack of proper housing, built shacks and turned areas into **shanty towns**, which are known as **bidonvilles** in France.

## Fringe development on the outskirts of Paris

- In Paris, due to serious housing problems, five new towns were built within a radius of 25 kilometres of the city.
- Up to 200,000 people currently live in new towns such as Cergy-Pontoise.
- New housing and industry were encouraged to set up in these towns and in growth centres such as Roissy, outside the city.

## The impact of management strategies

- The changes to the inner city of Paris were relatively successful; particularly the regeneration schemes. Traffic problems were reduced by the traffic-management schemes.
- The strategies were less successful in the ghetto areas. These areas are inhabited mainly by immigrants and their families from varied religions and ethnicities, which led to continual conflict, for example between those of Islamic and Jewish faiths.
- The recent terrorist attacks in Paris in January 2015 by Islamic fundamentalists had the effect of bringing these groups closer together in a show of unity in support of the victims.
- It remains to be seen whether this will continue in the years to come.

### Quick test ?

List the main reasons for changes to low-cost housing areas in a developed world city.

What are shanty towns?

**Example**

For a developed-world city, explain why there was a need for housing management. **8 marks**

### Sample answer

*In Glasgow, in the inner city, tenements in areas such as the Gorbals, were left decaying and derelict for many years because the city council couldn't afford to demolish or refurbish them. (✓) They were dangerous to live in and attracted crime and vandalism. (✓) They were cheaply built for Irish immigrants working in industry in Glasgow. (✓) During the 1960s the people who lived in these tenements were moved to new council owned housing estates on the outskirts of Glasgow such as Castlemilk, Easterhouse and Drumchapel. Other tenants moved to new towns such as East Kilbride and Cumbernauld.*

*The old tenements were demolished and some were replaced by new council houses such as New Gorbals and the Glasgow East Redevelopment (GEAR). Later, under new government schemes, some of these houses were bought by their tenants at a subsidised price.*

### Comment and marks

The answer shows some knowledge of the city that was studied. The opening three sentences give reasons why change was needed. The rest of the answer is irrelevant as it does not give any further reasons for change being needed but only describes the changes made. This answer would score **3 marks out of 8**.

# Change in developing-world cities

## The need for managed change and its impact: housing

### The causes for change to housing in developing-world cities

- The rapid increases in city populations due to in-migration from poorer rural areas and natural population increase due to high birth rates, lowering infant mortality rates and increasing life expectancy.
- Natural population increase is a result of improved economic development and improved standards of living and health care in the developing world.

*Key point*

With reference to a city you have studied, you should be able to explain change in cities in the developing world, relating to housing and transport, the management strategies adopted to implement these changes and the impact of the changes. (See Transport in developed- and developing-world cities on page 77.)

However, these rising populations have led to many social and economic problems, including:

- A lack of housing, which often results in homelessness, with families either living on the streets, in squatter areas or in shanty towns with a lack of proper facilities such as clean water, electricity and sewage disposal schemes. This leads to health and disease problems such as typhoid and cholera.
- A lack of industry affects employment opportunities and causes problems of low income and poor standards of living, including lack of food and poor education and health care.
- Infant mortality rates increase and life expectancy decreases.

### Management strategies adopted to implement change in developing-world cities

- Strategies include the cities benefiting from international aid schemes that provide money and medical and educational services for local communities. Aid schemes include funds from organisations such as United Nations sub-agencies, for example the United Nations Educational, Scientific and Cultural Organization (UNESCO), the World Health Organization (WHO) and the World Bank.
- Involvement in primary health-care schemes and local **self-help schemes**, particularly in shanty towns, are various ways in which social, economic and environmental problems are tackled.
- Success rates vary from country to country and city to city within the developing world.

For a developing-world city you have studied, you should be able to explain management strategies adopted to change housing quality, to improve poorer residential areas and to comment on the effectiveness of these efforts.

## The need for managed change and its impact: housing quality

### Explanation

- Housing quality varies due to lack of investment.
- Many houses lack basic amenities, particularly a clean water supply.
- Conditions may be very unhygienic and unhealthy.
- Some areas may be inhabited by squatters.
- Housing consists of poorly built wooden shacks that often house complete families in one room.
- These houses are rent free.
- There are few roads and often no electricity or access to waste disposal.
- There is often raw sewage in the streets.

### Improvement efforts

- Improvements to water supplies in many residential areas have been treated as a priority.
- Self-help schemes to improve local infrastructure using local labour have been financially supported by city authorities.
- Materials have been supplied to help improve the basic fabric of the poorest-quality houses.

- For example, in São Paulo (Brazil), efforts have been made to rehouse people in new-town areas. Financial incentives have been offered to residents to move to other parts of the city.

# The need for managed change and its impact: shanty towns

## Location and problems

- Shanty towns are usually found on the outskirts of cities in poorer areas, such as areas liable to flooding.
- The areas have developed due to homeless immigrants and high birth rates, which have caused rapid population increases.
- Houses lack sanitation, clean water, electricity and cooking facilities and may be constructed of the most basic materials, for example wood.
- The shanties lack services and proper roads and are usually located along major routes into the city.
- People may have to walk long distances to get into the city.
- Some people prefer to live in shanty towns as opposed to the insecurity of life on the streets.
- They provide contact with the extended family and may be a stepping stone to more permanent accommodation.
- Known as **favelas** in São Paulo, they consist of vast areas of poor, basic housing.

*Key point*

Referring to a developing-world city you have studied, you should be able to explain the location and development of shanty towns, their associated problems and management strategies adopted to improve conditions within them.

## Strategies adopted to manage these problems in São Paulo

- City authorities have introduced planning legislation to build satellite towns.
- Authorities make provision for sites and services for self-help housing away from the shanty towns.
- There is also provision of loans and grants to help build new houses or improve existing ones.
- There are improvements to local infrastructure, for example power, water, sewage and roads, and selective clearance of the worst areas.
- Authorities are involved in the building of new housing estates with low-cost housing and basic facilities.
- There has been a massive increase in police activity to tackle major crime issues such as drug trafficking and violent crime. Locally, this intervention is called **pacification of the favelas**.

## Negative impacts of these strategies

- Costs have increased.
- Inhabitants have been reluctant to move.
- Some strategies have caused a breakdown of community and family units.
- Skilled training is needed but doesn't necessarily exist.
- Rents can be too high in new housing areas.
- Populations within shanty towns increase as improvements are introduced.

# Case study 3: a shanty town – Rocinha, Rio de Janeiro

Rocinha is the largest favela in Brazil, located at the southern edge of Rio de Janeiro. It is built on a steep hillside overlooking the city (see Figure 2.12 on the next page). The land is very unstable and liable to landslides. The dwellings are extremely unsightly and cause visual pollution to the nearby main tourist beach.

- The settlements are usually very overcrowded.
- Rocinha is home to between 70,000 and 180,000 people. There is a lack of services like schools and hospitals, so it is hard to break the cycle of poverty.
- The authorities in Rio de Janeiro have taken a number of steps to reduce problems in favelas.
- They have set up self-help schemes. The local authority provided local residents with the materials needed to construct permanent accommodation, for example breeze blocks and cement.
- The local residents provided the labour. The money saved can be spent on providing basic amenities such as electricity, water and perhaps a school. Today, almost all the houses in Rocinha are made from cement and brick.
- Roads have been built to improve communications into Rocinha. Parts of Rocinha are accessible by bus.
- The authorities have set up a small number of health clinics that provide basic health care and at least one gynaecologist and two paediatricians.
- A majority of houses have running water and 99 per cent have electricity, although many obtain power illegally by tapping into power lines.
- The permanence of Rocinha is now accepted by the Brazilian government, even though the land was initially occupied illegally.
- In order to clear the area of crimes such as drug trafficking to make Rocinha safer for its inhabitants, a process known as pacification was adopted. This involved large numbers of police patrolling the favela to root out criminals.
- Basic infrastructure has been provided and Rocinha has become an area of permanent settlement, integrated into the formal city through the legalisation of some land holdings.

*Key points* 

- Population: 180,000
- One of the biggest slums in Rio.
- Housing was originally made from cardboard, mud and tin sheets; most are now made of brick and cement.
- Open sewers and toilets are seen everywhere within the favela.
- 99 per cent of houses have access to electricity, but most of this is via illegal 'tapping'.
- There are not many maintained roads.

**Figure 2.12** A typical shanty town in Rio de Janeiro (Brazil)

# Case study 4: Mumbai (formerly Bombay), India

## Reasons that change was necessary in Mumbai

- Mumbai is India's largest city. A rapid increase in city population led to massive demand for housing.
- Natural growth and the mass influx of people from upcountry and from neighbouring countries like India, Bangladesh, Sri Lanka, Burma and Afghanistan created this rapid increase.
- Many people move from the countryside to the city, swelling its population. They move to the city in search of new opportunities, employment and better health care.
- As most of the migrants are young they are more likely to produce children of their own, further increasing the city's population.
- More health care is available in the city, and better health care reduces infant mortality and increases life expectancy, further increasing the urban population.
- Overseas aid is more available in the cities, thus attracting migrants to the cities.
- One of the main difficulties that has resulted from the rapid growth of Mumbai is the increase in shanty towns.
- Shanty towns are unplanned and have incomplete water and sewage systems and drainage systems.

- Slum dwellers make up 60 per cent of Mumbai's population: approximately 7 million people. Slums are built on any space available, even on the sides of roads.
- Infant mortality is as high as it is in rural India, where there are no amenities. General hospitals in the Greater Mumbai region are overcrowded and under-resourced.
- As a result of Mumbai's size and high growth rate, urban sprawl, **traffic congestion**, inadequate sanitation and pollution pose serious threats to the quality of life in the city.
- Vehicle exhausts and industrial emissions, for example, contribute to serious air pollution, which is reflected in a high incidence of chronic respiratory problems among the population. In fact, most people rely on private doctors, many of whom do not have any qualifications or official training.
- There is high unemployment, made worse by large numbers of people arriving each day.
- There is a growth in trading black-market goods.
- Drugs, crime and prostitution are common in parts of Mumbai as the urban poor try to survive.
- Wages are poor as there is such a large labour force looking for work.

## Strategies adopted to manage change in Mumbai

- Efforts to solve the problem of a lack of clean water supply to housing areas in the city involved the World Bank, which funded the development of 176 Primary Care Dispensaries.
- In 1976, the government passed the Urban Land Act, which was supposed to enlarge the area on which middle- and lower-class housing was to be built.
- In 1985, the government tried to rectify the problem of slum dwellings by passing the Slum Upgradation Project. It offered secure long-term legal plot tenure to slum households on the basis that they would invest in their housing.
- By giving people an interest in their housing and by guaranteeing home ownership, the government hoped to get rid of the slums.
- The government has tried to improve public transport to reduce congestion and pollution.
- The government of Maharashtra is planning to introduce a congestion charge in Mumbai to relieve the traffic problems that plague the streets.
- Billboards are being used to try to encourage people to leave their cars at home.
- Car-pooling is also being suggested.

## The impact of these strategies on the city

- The World Bank Primary Care Dispensaries are supplying some medical care facilities, but these are underused and the water supplies needed are unreliable.
- There is always too much or too little water; when monsoon season hits, some slums are submerged knee-deep in water.

- The Urban Land Act has not been successful as the Act has been used not to build affordable housing for the slum dwellers but to build more upper-class housing and to keep hold of wealthy neighbourhoods, which has worsened the slum problem.
- The Slum Upgradation Project has been partially successful but unfortunately the programme targeted only 10 to 12 per cent of the slum population – those who were capable of upgrading their homes. It disregarded those who did not have homes at all.
- By relocating more than 6000 illegal slum dwellings that encroached on land adjacent to the railway tracks, Indian Railways increased its service dependability, speed and safety.

## Example

'A shanty town is a group of unplanned shelters constructed from cheap or waste materials (such as cardboard, wood and cloth). Shanty towns are commonly located on the outskirts of cities in poor countries, or within large cities on derelict land or near rubbish tips.'

For any city you have studied in the developing world, discuss the economic, social and environmental problems created by shanty towns. **6 marks**

### Sample answer

*Agua Fonte is a favela in São Paulo. In São Paulo shanty towns have no sanitation or sewage removal scheme, (✓) raw sewage causes disease such as typhoid and cholera, which spreads quickly through closely packed houses in favelas. (✓) The people get these diseases through unclean drinking water and vermin. The favelas are built on hillsides which could collapse after heavy rainfall. (✓) They are illegally built. (✓) Crime rates are high due to poverty, drug trafficking, prostitution etc. (✓) There are no buses from these areas into centre for workers. There is a high rate of unemployment. (✓)*

### Comment and marks

This is a good answer, showing good knowledge of the case study. The answer refers to the main features of the shanty town and its problems and gives good examples, such as 'favelas' and 'Agua Fonte'. It scores a total of **6 marks out of 6**.

# Transport in developed-and developing-world cities

The need for change in transport in cities throughout the world is due primarily to vast increases in traffic, which have led to major traffic congestion problems.

Traffic congestion is probably the greatest transport problem found in most cities in the world, especially in developed-world cities.

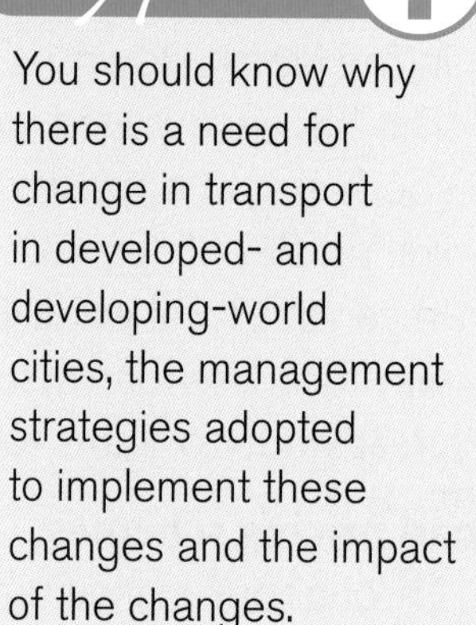


## Transport issues in developed-world cities

Traffic is one of the most difficult problems that cities have had to deal with in recent years.

### Reasons for increased traffic congestion in developed-world cities

- Commuter traffic from rural/urban fringe areas surrounding developed-world cities increases the volume of vehicles on the roads.
- The volume of traffic has increased tremendously, with huge numbers of private and commercial vehicles entering and leaving city areas throughout the day.
- The problem is usually at its greatest during the morning and evening rush hours. However, weekend traffic is often just as bad.
- This has led to problems of congestion, accidents, damage to roads, high costs of road maintenance and disruption to the life and work of the city.

### Management strategies

**Ring-road systems:**

Figure 2.13 shows the orbital road system around London, which is based on the M25 motorway.

- This road system provides a convenient way around London and links with the major motorways and Channel ports.
- It has eased traffic congestion within the city.
- There are similar ring-road systems operating throughout many parts of the developed world.
- All of these are aimed at reducing the volume of traffic travelling in and out of major city areas.
- Traffic using these systems can bypass city areas, thus reducing problems of environmental pollution, road maintenance, traffic congestion and road accidents in inner-city areas.

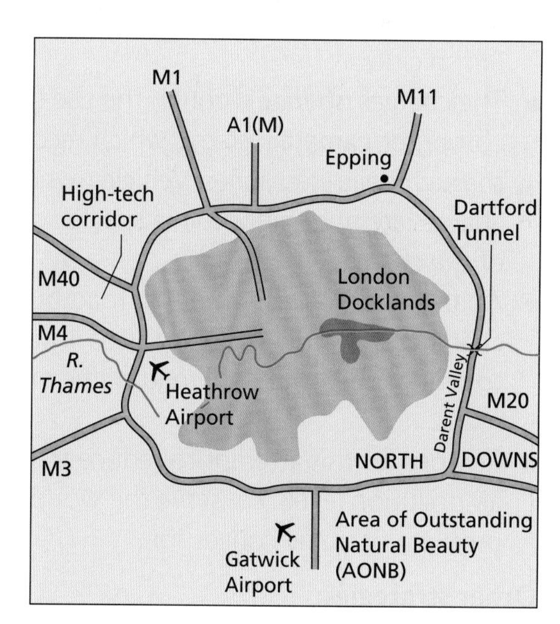


**Figure 2.13** The orbital road system: London

However, as well as helping to reduce traffic problems in cities, this strategy has also created problems for the surrounding areas:

- Property values near the London orbital have increased, since industry and commuters see the advantages of living close to motorways.
- The London orbital put development pressure on land in the **green belt** and destroyed many attractive areas of countryside, for example in the North Downs.
- This has encouraged industry and commerce to invest more in the south east as opposed to the more economically depressed areas of the north of England.
- In general, local wildlife habitats have often been destroyed to make way for these road systems and local pollution levels have increased greatly due to traffic fumes.

**Road pricing schemes:**

- Throughout many European countries, road users are charged a toll on major transport routes, especially motorways.
- These charges are directly related to the distance travelled. There are different rates for different types of vehicles: private cars, road haulage vehicles and tourist buses all pay different amounts.
- These charges contribute to the cost of building and maintaining the motorways.
- Apart from toll charges on bridges such as the Forth Road Bridge in Scotland and the Severn Bridge connecting the west of England to Wales; tunnels such as the Dartford Tunnel in London, and the M6 toll road, there are no charges, as yet, for those using Britain's motorways.
- There are, however, charges for those driving on inner-city roads in London, and there are government and local authority plans to charge inner-city road users, both private and commerical, in other major British cities.
- The charges operate through the use of highly technological computer/camera systems, which record and log vehicles passing through delimited zones. Vehicle owners are sent a bill, and if road charges are not paid within a specific time limit, fines are issued on top of the charges.
- Apart from raising revenue for road building and maintenance, charging drivers in inner-city areas is part of a government strategy to reduce the volume of road traffic on Britain's and other European roads.
- Reasons for this attempt to reduce traffic include the need to reduce air pollution levels, conserve the world's oil supply and to improve the problem of traffic congestion.

**Other strategies:**

When attempting questions on this topic, strategies that should be mentioned in good answers would include:

- Changes to **road systems**, such as ring roads, one-way systems, use of 'bus only' lanes, contra-flow systems, widening of roads, building bypasses.

- Major streets have become either **pedestrianised** or have **one-way** traffic flows.
- **Parking** restrictions are also used to prevent parked cars blocking main routes. These are enforced by police and traffic wardens.
- The provision of **alternative means of transport** to encourage drivers to leave their cars outside the city.
- **Park and ride schemes**, whereby car users park their cars outside the city and use public transport to complete their journey into the city centre.
- Cheap public transport on **buses**.
- Metro or **underground rail systems** with road/rail linkages have proved very successful.
- Suggestions that drivers commuting to the cities should be charged additional **taxes** on their cars in the form of tolls and charges for private parking places.

**Problems created and the negative impact of these strategies:**

- Preventing people from coming into the city can create problems, especially for shopkeepers and other businesses that rely on customers for their existence.
- Offering alternatives such as out-of-town shopping centres may help solve traffic problems but providing these centres often leads to the decline of the central business district (CBD), from which a large proportion of local government finance is obtained.
- Authorities have to balance their solutions very carefully, and national government policies recognise this. New proposed legislation in Britain, for example, includes imposing tolls on cars entering the city, increasing fuel prices and upgrading railways and other public transport services.

# Case study: Glasgow

- In Glasgow, the main routes meet in the centre of the city, making Glasgow's main business area very accessible to thousands of people.
- The M8 motorway runs through the city centre, giving access to thousands more from across the central belt of Scotland.
- The main train and bus stations are found in the city centre, including Queen Street Station and Central Station, which allow shoppers and workers easy access to the city's central areas.
- Narrow streets in the centre are laid out in a gridiron pattern and cannot cope with the present-day volume of traffic. Therefore streets such as Hope Street, Sauchiehall Street and Renfield Street are now one way to ease traffic congestion.
- Multi-storey car parks were built at the St Enoch Centre and Buchanan Galleries shopping centres to stop cars parking on the streets and to keep traffic flowing.

# Transport issues in developing-world cities

## The need for change to transport systems in these cities

The need for change to transport systems in these cities is due primarily to problems closely linked to vast increases in city populations, where people move to inner-city areas for employment.

- Problems are similar to those in developed-world cities but tend to be even worse in many developing-world cities, such as in São Paulo, Brazil, due to the lack of money to pay for the ever-increasing costs of providing a transport infrastructure that works.
- Many developing countries inherited a transport network system from their previous colonial rulers.
- Emphasis in the past was to build large ports and high-grade roads and railways to exploit the resources of these countries.
- Such development left little money to improve access to and from the poorest parts of these countries.
- Poor-quality roads within inner-city areas are unable to cope with the volume of traffic.
- There is a lack of adequate public transport services in most cities.
- Problems of traffic congestion are due to factors such as quality of roads leading to city centres, huge volumes of commuter traffic at peak times, lack of or cost of public transport services, parking on main roads and lack of parking facilities within city centres.
- Developing countries need efficient transport networks to aid their development. Present networks in Africa and South America encourage a **polarisation of resources**. This means that more and more of the wealth of these countries is concentrated in a very small number of areas, usually in the large cities. This trend has resulted in many rural areas in these countries being deprived, with high levels of unemployment and poverty.

**Strategies adopted in developing-world cities:**

- An increase in the provision of public transport including buses and trains.
- Bicycles are a popular means of transport in many countries such as India and in south-east Asia and South America.
- A lack of money has inhibited the construction of new roads and a more efficient transport infrastructure in many developing countries.

## The need for management of recent urban change in developed-world cities

The need for management is a direct result of a number of issues affecting transport in developing-world cities.

Essentially these issues are similar to some of the issues faced by developed-world cities. These issues include:

**Population growth of cities:**

- Most of the world's population growth over the next 30 years will occur in cities and towns of developing countries.
- Population growth and economic development have caused rapid increases in the number of motorised vehicles in cities.

**Traffic congestion:**

- With rapid urbanisation and economic growth, motorisation has increased in cities in developing countries, for example in the Asian region the number of motor vehicles per 1000 people has more than trebled in the last 30 years.
- Road engineering has traditionally been to 'build your way out of congestion'. This identifies the problem of congestion as being a lack of roads and the solution as improving traffic flow and ignores other, more complex, problems of travel-demand management and the negative side effects of building more roads. In the developing world, the trend is still this expansion of infrastructure for private motor vehicles and not the improvement of public transport.
- Increasing road construction has failed to cope with the ever-increasing demand from rapid motorisation, resulting in a vicious cycle.

**Lack of finance:**

- Developing countries lack the funds for major transport developments such as building new roads, infrastructure, off-street multi-storey car parks and monitoring systems and the funds to employ wardens to enforce the parking rules.
- Those seeking to obtain finance for transport improvements have to compete with demands for other city resources, such as housing, education, health services, administration, shops and business centres.
- Agencies such as the World Bank have provided grants, but they are often insufficient to make any real difference to the transport services in developing-world cities.
- Public-transport systems have deteriorated over the last two decades, often as a result of the move from state-controlled to market economies.
- In some developing-world countries, public transport has never been managed by public authorities, but by private operators.
- Motorcycles have become a popular option in cities where cars remain unaffordable for many.
- In Hanoi, for example, in the mid-1990s 61 per cent of trips were made by motorcycle, 30 per cent by bicycle and only 3 per cent each by bus and car.
- Since then, the numbers of motorbikes have continued to outstrip that of cars, and at an even faster rate.

**Narrow roads and streets:**

- In most developing-world cities, streets and roads are very narrow due to historical reasons and a lack of urban planning. When built, these routes were not designed to cope with the volume of traffic – of all types – now seen, including motorised transport.
- Streets were designed to cater for more primitive types of transport, including horse/oxen-drawn carts, pedestrians and more ethnic forms of transport, such as rickshaws.
- As some of these countries developed, they saw a massive increase in motor vehicles, including cars, lorries, buses, motorcycles and scooters, as prosperity increased.
- Efforts to improve and widen roads have been hampered by financial restrictions.

**On-street parking:**

- On-street parking increases road congestion due to vehicles being parked on public roads and streets, thereby narrowing the routes and leading to traffic jams.
- Developing-world cities often lack the resource of manpower, that is, police and traffic wardens, to monitor and penalise illegal parking.

**Public-health issues:**

- Transport is responsible for some of the most serious environmental hazards and health risks faced by many developing cities.
- However, the health, environment and transport authorities in many developing countries do not attempt to deal with transport-related health risks as a priority.
- Healthy transport systems emphasise the efficient movement of people, not just of vehicles. In large cities, such systems generally include:
  - exclusive high-capacity networks and corridors for urban buses and trains
  - modern, high-quality networks for pedestrians and cyclists
  - integrated land-use planning for healthy urban spaces.

**Health risks associated with transport:**

- Urban air pollution, much of it from vehicle emissions, causes cardiovascular and respiratory diseases, among other illnesses, and kills some 1.2 million people annually around the world.
- Road-traffic injuries are responsible for another estimated 1.3 million deaths; pedestrians and cyclists are among the groups most at risk.
- Many injuries result from poor transport and land-use designs, and particularly from the lack of safe space for non-motorised transport such as pedestrians and bikes.

## Management strategies

**Improvements to public transport services:**

- The public benefits of investment in urban public transport systems outweigh the costs.
- Public transport works best if it is possible to travel from any point in a city to any other point.

**The use of alternative means of transport such as metro systems:**

- Urban transport in the newly industrialised countries in East Asia is dominated by the problems of the primate cities.
- The main ones, Seoul, Bangkok, Manila, Jakarta and Kuala Lumpur, have historically all been dominated by road transport.
- Most have constructed urban expressways – but all still have a great deal of congestion and pollution.
- The urban rail transport systems in these cities, which are intended to relieve the congestion and pollution problems, vary greatly in development and efficiency.
- With the exception of the Korean cities, suburban railways are usually poorly operated by the national rail companies and make little contribution to the urban transport network.

**Strategies related to health issues:**

- A healthy urban transport system must include substantial allowances for walking and cycling.
- These modes of transport remain important in the developing cities of many regions. Cities in the People's Republic of China have the highest rates of non-motorised transport (NMT).

Other strategies include methods known as AVOID, SHIFT, IMPROVE.

**Avoid:**

- This first strategy aims to avoid unnecessary travel and to reduce trip distances. It includes integrated land-use and transport planning and mixed-use development, as well as using information and communications technology (ICT) to reduce numbers of trips.
- Mixed-use development includes locating home, work and shopping areas close to each other, to increase accessibility and reduce trip distances and times.
- ICT can replace many activities that previously required travel.

**Shift:**

- This strategy involves more sustainable transport modes.
- It encourages people who use motorised vehicles to take more public and non-motorised transport.
- It involves travel-demand management measures and better development of inter-city transport.
- It aims to persuade people who already use public and non-motorised transport to continue to use these sustainable modes.
- The 'shift' approach is the second-best means of delivering sustainable urban transport, after 'avoid', if implemented properly.
- If 'avoid' and 'shift' strategies are applied in a city, most of the hard work has been done, though improvements can still be made.

**Improve:**

- This third strategy focuses on improvements to transport practices and technologies. It is a more technological approach to tackling urban transport problems.
- Its measures include improving fuel quality, vehicle fuel efficiency, emissions, inspections and maintenance and moving to 'intelligent transportation systems' that use ICT to improve transport management.

**Efforts to improve the quality of roads:**

- Road building is expensive. Much of the urban transport investment in developing countries in the 1990s was focused on road improvements. A recent analysis of 100 cities worldwide found that:
  - among cities in developing regions, only Latin American cities invested a greater proportion on urban public transport systems than urban road development
  - in three of the People's Republic of China's five largest cities, spending on road development and maintenance was 3.7 times greater than spending on public transport.

**Impacts of 'avoid, shift, improve':**

- The Bangkok 2020 Declaration states that there should be a fourth strategy that emphasises a 'people first' approach: a more humane sustainable urban transport policy that improves safety in urban transport systems, delivers health benefits and reduces air pollution and noise.

## Key words and associated terms

**Dereliction and decay:** A reference to old buildings such as factories and houses that, through age and wear and tear, are no longer usable and have been abandoned.

**Green belts:** Areas surrounding cities and towns in which laws control developments such as housing and industry in order to protect the countryside.

**Inner city:** The area near the centre that contains the CBD, the older manufacturing zone and a zone of low-cost housing.

**Park and ride schemes:** An attempt to reduce traffic congestion by encouraging people to park their cars outside city limits and use public transport for the remainder of their journey into the city.

**Pedestrianised zones:** Traffic-free areas within the city centre where people can shop and walk along streets where traffic is forbidden to enter.

**Renewal and regeneration:** The processes by which older areas of cities are demolished and replaced by new buildings, which often have totally different functions from the original building or area.

**Ring roads:** Roads built specifically to take traffic away from the city centre and to help solve the problem of congestion.

**Rural/urban fringe:** The area at the edge of the city where the city limits meet the countryside. It is attractive to developers for a variety of purposes, such as business parks, leisure and recreational parks, out-of-town shopping centres and motorways. Development may be restricted by legislation such as green-belt laws.

**Self-help schemes:** Projects in which local people become involved to improve their living conditions.

**Shanty towns:** Areas found mainly in and around cities in developing countries, usually consisting of temporary and makeshift accommodation made from materials such as wood and corrugated iron. In South American countries they are called 'favelas' and in India they are known as 'bustees'.

**Suburbs:** Housing zones on the outskirts of towns away from the busy central inner city.

**Traffic congestion:** The heavy build-up of traffic along major routes and within city centres that causes great problems of cost and pollution for many cities, in both developed- and developing-world cities.

# Section 3 Global issues

## Chapter 3.1 River basin management

### Sites for dams and reservoirs

The physical and human factors that influence the choice of site for dams and reservoirs include:

- the need for solid foundations for a dam
- the need for a narrow cross-section to reduce the dam length
- a large deep valley to flood behind the dam
- a sufficient water flow from the catchment area
- local evaporation rates
- the permeability of underlying rock
- the amount of farmland that would be flooded
- the number of settlements that would be flooded
- local population distribution
- the distance from urban or farming areas for hydroelectric power (HEP) or irrigation.

*Key point*

You should know the physical and human factors that influence the choice of site for dams and reservoirs.

### The size and shape of the catchment area

- Depending on the basin selected for study, the size and shape of the catchment area can vary from a basin that is located completely within the boundaries of a single country to one that flows through several states or countries.
- Examples of basins located completely within a country include the Amazon basin in Brazil, the Hwang Ho basin in China and the Indus basin in India.
- Examples of basins flowing through several states/countries include the Nile in North Africa, the Mississippi and Colorado basins in the USA, the Ganges basin in India and Bangladesh, and the Murray Darling basin in Australia.
- You should know the overall size of your selected basin in square kilometres and the area covered. If it flows through different states or countries you should be able to provide examples of these.
- You should be able to mention the main **tributaries** found in the basin.

*Key points*

Referring to a selected river basin, you should be able to describe and explain:

* the size and shape of the catchment area
* rainfall distribution and reliability
* surface features
* rock type.

### Rainfall distribution and reliability

- You should have some general knowledge of the main features of the climate(s) within your selected **river basin**.
- You should also be able to discuss rainfall distribution throughout the year, referring to any periods where rainfall is excessive or unreliable.

- If the basin covers several climatic areas you should be able to refer to this.
- You should note the effect of the climates on river flow.

## Surface features

- Rivers can flow through a variety of terrains.
- Basins such as the Nile, Ganges and Amazon contain a variety of landscapes, such as mountains and wide flat floodplains that terminate in wide deltas.
- These landscapes contain surface features that can interrupt and affect river flow, for example waterfalls, rapids, gorges and rainforested areas.
- Some of these surface features may offer opportunities for **water control projects** and the construction of multi-purpose dams for water and electricity supply.

## Rock type

- The variety of rock types will change considerably during the course of the river, from source to mouth.
- These rock types may have an impact on the site chosen for the construction of dams and reservoirs.
- You should have some general knowledge of the variety of rock types present within your selected basin and be able to comment on their impact on river flow and suitability for water control projects.

# Water-management schemes

When answering a question on this topic, your answer will depend on the river scheme chosen, but may refer to the following:

**Key point**

For a water-management scheme you have studied, you should be able to explain the need for this project and the social, economic and environmental impact it has had on the basin. You should also be able to assess the success of the scheme.

## Economic advantages

- Improved yields in farming.
- Hydroelectric power, which is helping to create industrial development.
- More water for industry.
- Improved navigation channels.
- Control over flooding.

## Economic disadvantages

- The new scheme could involve huge expenditure.
- It may depend on foreign finance, resulting in increased debt.
- More money may be needed for fertilisers and compensation to land owners.

## Social advantages

- The greater population is helped by increased food supply.
- There is less disease and poor health due to a better, safer water supply and more food becoming available.
- There may be more recreational opportunities.
- There is more widespread availability of electricity.

## Social disadvantages

There will be forced removal of people from valley sites and increased incidences of water-related diseases such as bilharzia/schistosomiasis and river blindness.

## Environmental advantages

- There is an increase in fresh water supply.
- There are improvements in sanitation and health.
- There are also improvements in scenic value.
- There is the bonus of flood control.

## Environmental disadvantages

- Water pollution and industrial pollution increases.
- There is an increase in **silting** of reservoirs.
- There are also increased **salinity** rates further downstream.
- There may be possible flooding of historical sites.

## Political consequences

Political problems depend on the river basin you have chosen to study, but might include:

- water control or dependence on neighbours upstream
- complex legislation over appropriate water sharing by different states or countries
- reduced flow and increased salinity in some areas
- shared costs between states or different authorities or problems over the allocation of costs
- increased pollution across borders, resulting in problems with allocating appropriate costs for cleaning the river.

You should know the political problems that may have resulted from management project(s) and how political considerations might hinder the project(s).

## The hydrological cycle of the basin

- There is increased evaporation from the larger surface area of massive reservoirs.
- There is an impact on the local climate.
- There is less water flowing below dams.
- Rivers are often diverted.
- Water table levels change.
- **Infiltration** rates are affected by water held in reservoirs or irrigation channels.

Hints & tips

*In the examination you may be given a range of maps, climate graphs, hydrographs and so on for any river basin, for which you should be able to explain why there is a water supply problem and why a water storage scheme is needed.*
*Alternatively, you can discuss a river basin project that you have studied.*

# Case study 1: the Nile basin

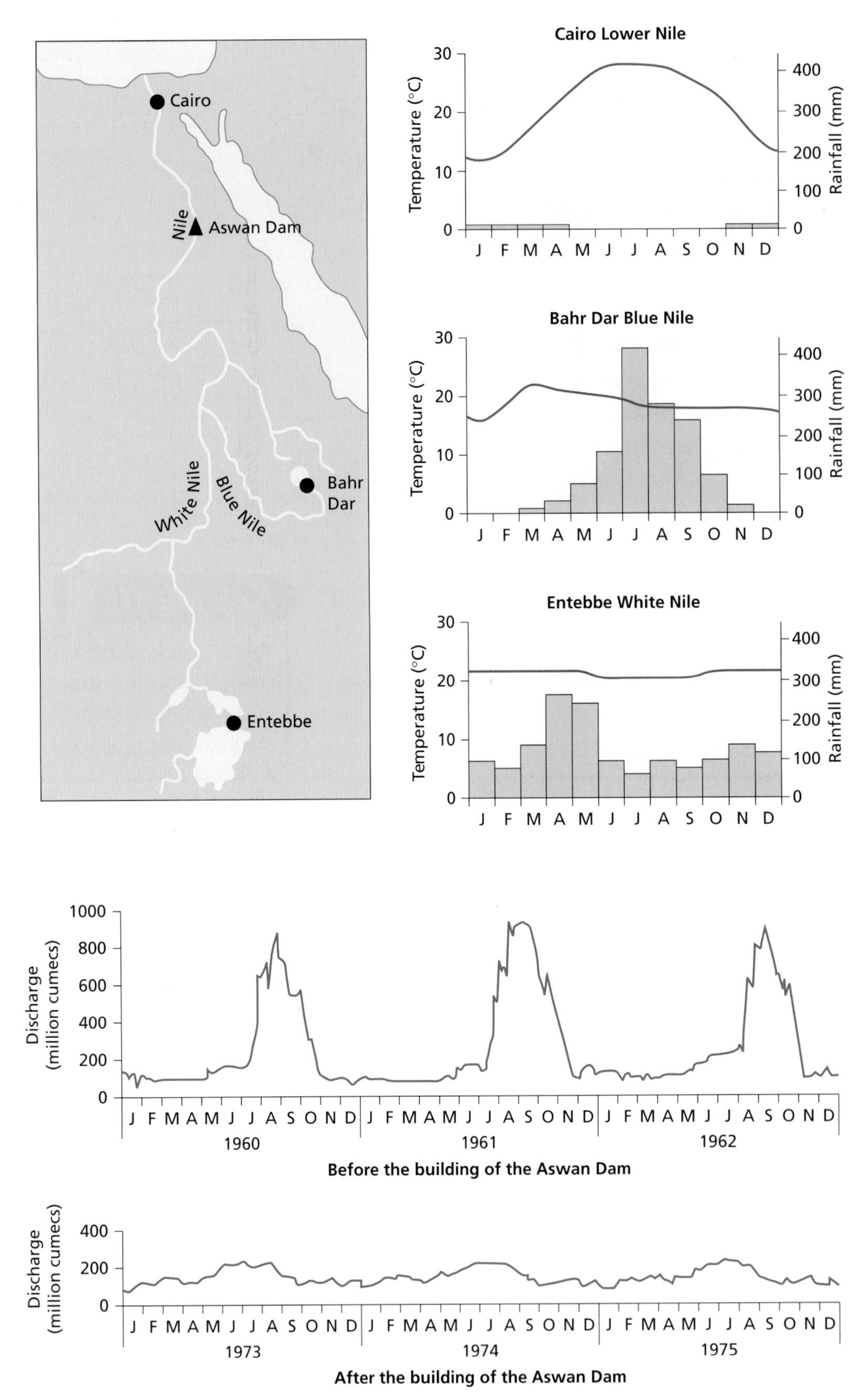


**Figure 3.1** Nile basin data

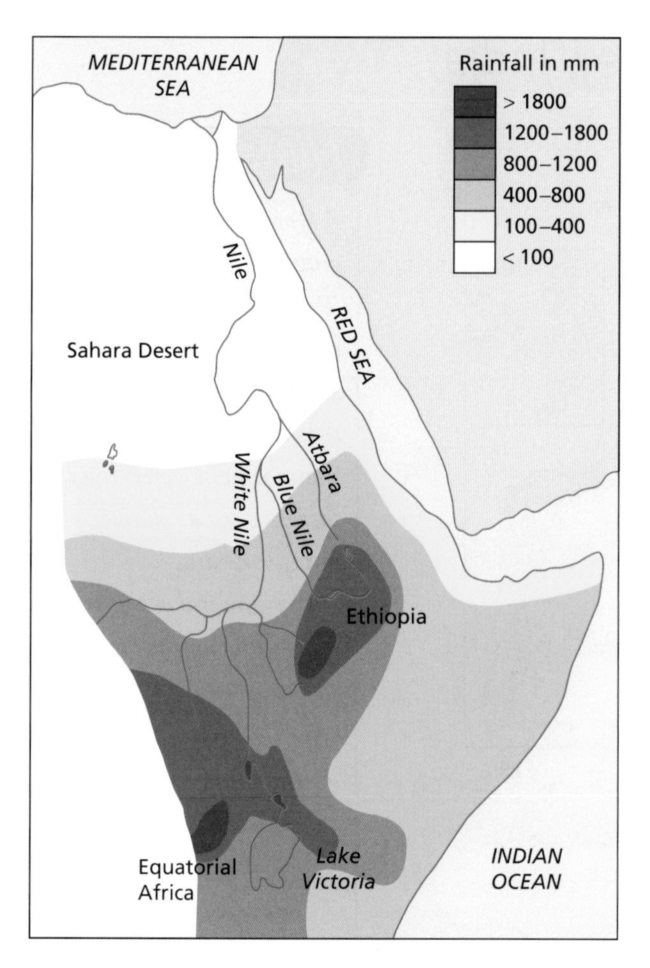


Figure 3.2 The Nile basin

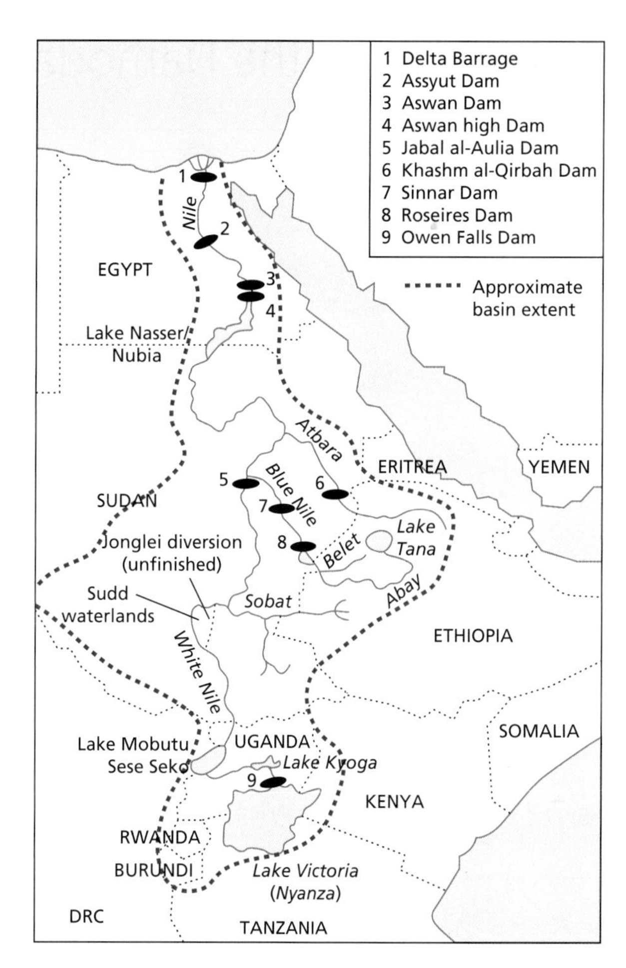


Figure 3.3 Rainfall in the Nile basin

For the figures given for the Nile Valley (see Figure 3.1), you should note:

- the wide **seasonal fluctuations** in discharge levels
- the steady flow of 100–500 million cumecs between November and May, rising to 200 million cumecs in June
- a rapid rise in July to a peak discharge of 900 million cumecs in late August/early September, with an equally sharp decline in October.

## Explanation

- The July to September surge is due to the pronounced seasonal regime of the Blue Nile's catchment area in the Ethiopian Highlands (see the climate graph for Bahr Dar in Figure 3.1).
- The more regular flow of the White Nile, with its source in East Africa (Lake Victoria), helps compensate for the dry season in Ethiopia (see Figure 3.2 above).
- This maintains water levels at a steady level for the remainder of the year.

## Change

- The flow of the river has been even and regular since the Aswan Dam was built.
- Extremes giving rise to annual floods no longer occur.
- Despite slight fluctuations, maximum discharge rarely exceeds 250 million cumecs.
- The dam clearly controls river flow.

# Case study 2: the Damodar Valley, India

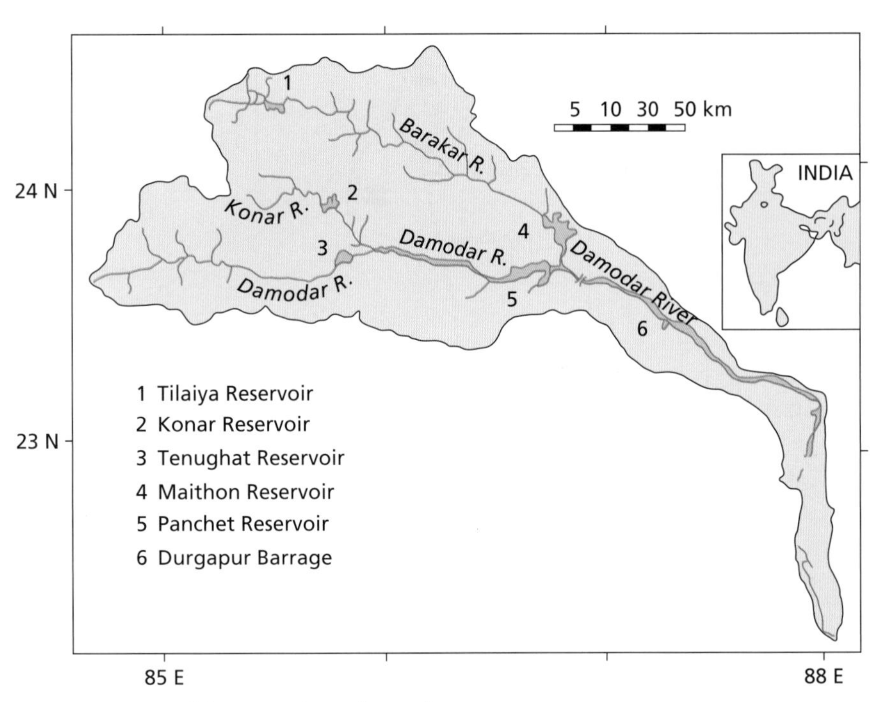


**Figure 3.4** Map of the Damodar Valley, India

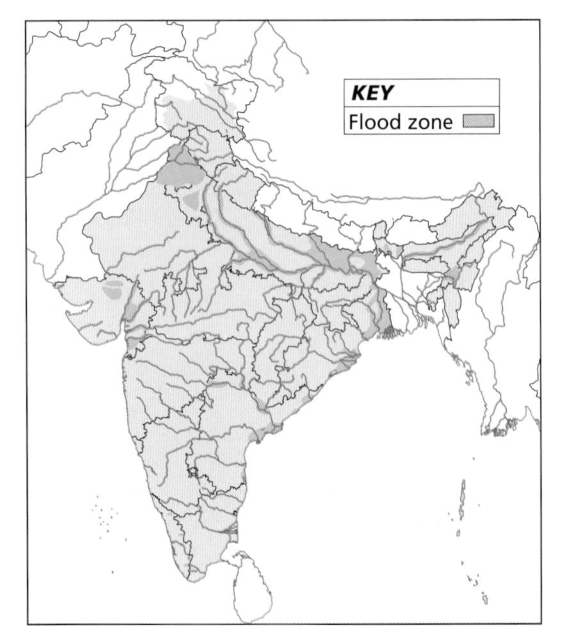


**Figure 3.5** Flood zones, India

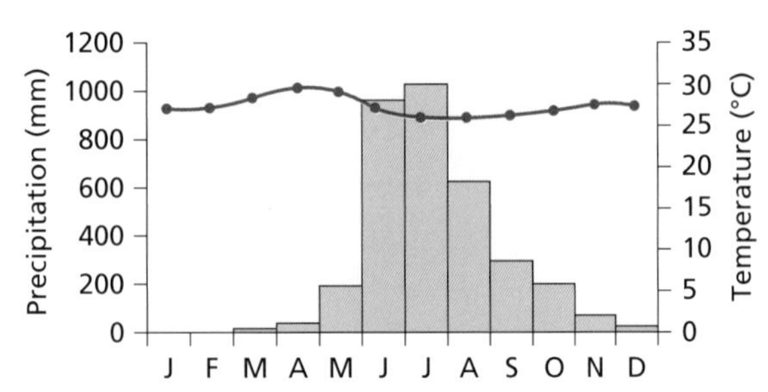


**Figure 3.6** Climate graph of Panchet, India

## Hints & tips

*A typical examination question might be: 'Referring to a named river basin that you have studied, explain why there was a need for river basin management, referring to factors such as climate, landforms and environmental problems.' Depending on the basin chosen, your answer may refer to:*

- ✓ *the rainfall pattern, especially periods of low rainfall*
- ✓ *any seasonal droughts*
- ✓ *any navigation problems due to physical features of the river valley*
- ✓ *the need for water transfer to areas of deficit*
- ✓ *the irrigation potential and flood control during high rain seasons*
- ✓ *any hydroelectric power potential and the suitability of underlying geology (impermeable rock) for water storage schemes and dam construction.*

## Reasons for the need for the project

- Rainfall in the Damodar Valley is seasonal.
- At some points in the year there is an excess of water, for example over 2000 mm in the months of June and July. This led to the need for flood control.
- Figure 3.5 shows that the Damodar river valley is in the area of India that is prone to flooding.
- However, there is very little rainfall between the months of December and March, which causes drought. Since there is an excess of water in the wet season, this could be stored and used in the dry months.
- The Damodar River has many tributaries so there is a high **drainage** density in the river valley, which increases the risk of flooding.
- The amount of water in the river valley can be variable and unpredictable; planning can help regulate the use and distribution of the water available.
- India's population is rising so there is a need for a constant supply of water for power, domestic use and industrial needs.
- Increasing population means more demand for food supplies, so farmers need a constant supply of water to irrigate their crops during the dry season.

## Needs for constructing the dam

- The dam should have solid foundations to support the great weight of the dam.
- There should be impermeable rock below the reservoir to prevent seepage and water loss.
- There should be enough rainfall/water in the river basin to supply the dams and associated reservoirs.
- The catchment area should have a sufficient flow of water to fill reservoirs.
- The dam should be built at the narrowest point across the river to reduce the length of the dam.
- There should be a large deep valley to flood behind the dam to allow a reservoir to form.
- If possible, the construction of dams should be avoided in earthquake-prone areas. The weight of the water stored in the dam can destabilise the underlying strata and cause an earthquake.
- The lower the temperature the better as this will reduce water loss through evapotranspiration. However, the temperature graph for Panchet (Figure 3.6) shows constantly high temperatures throughout the year, leading to high evaporation rates. Monthly temperatures average around 27° C and are highest when rainfall is lowest.

**Quick test ?**

What are the main physical and human features considered when choosing a site for dams and reservoirs?

## Impacts and consequences of the project

**Economic advantages:**

- A more reliable water supply increases crop production, which produces a surplus for sale and improves the economy of the country.
- The availability of cheap electricity has encouraged industry to move into the area, allowing the rich minerals of the Damodar Valley to be developed and removed.
- Industry has provided employment for local people, improving the local economy as well as developing the economy of India as a whole.
- The course and depth of the river are more constant, which improves transport and trade links.

**Economic disadvantages:**

- These schemes are very expensive to build and maintain.
- In developing countries, money for river schemes may need to be borrowed and could lead to debt.
- As the river does not flood on a regular basis, the silt from the flooded river no longer reaches the fields so there is a greater need for farmers to use fertilisers, which increases their costs.
- Roads and other infrastructure can be lost under newly formed reservoirs, thus disrupting communications.

**Social advantages:**

- The local populations have an improved water supply, which leads to better health and the reduction of water-borne diseases like cholera.
- As a result of water being available all year round, farmers can grow more crops, increasing food supply. This can sustain India's growing population.
- The dams provide water for domestic use all year round. The production of hydroelectric power has increased the supply of electricity to the local populations.
- The dams and reservoirs at Panchet and Mython are being developed for tourist resorts and sightseeing, providing opportunities for locals and visitors.

**Social disadvantages:**

- The local people are forced to leave their homes as large areas are flooded behind the dams to form reservoirs.
- They are resettled in areas that are less fertile and provide a poor livelihood.
- Water-related diseases such as malaria increase in irrigation channels.

**Environmental advantages:**

- The danger from flooding has been reduced and river flow is reliable throughout the year.
- An increased fresh water supply improves health and sanitation.
- Some people think that the dams and reservoirs are an attractive view.
- The reservoirs attract wildlife into the area.
- The frequent floods wash away the weeds that clogged the drainage systems, especially in the lower channels of the river.

**Environmental disadvantages:**

- High embankments are put in place to control river flow but villages such as Dadpur are constantly under threat of inundation from breaches in these embankments, which could result in the whole village being washed away.
- The frequency of floods has been reduced but they have become much more unpredictable.
- Large amounts of sand and silt are washed down the river, causing soil erosion and silting up of the reservoirs.
- The traditional river floods carried large quantities of silt, which created fertile growing conditions.
- The incidence of malaria has increased as the breeding grounds are no longer destroyed by the flooding river.
- There is increased water pollution due to all the new industries that have been attracted into the area.

# Case study 3: the Colorado River scheme

## The physical and human factors that influenced the choice of site for the dam

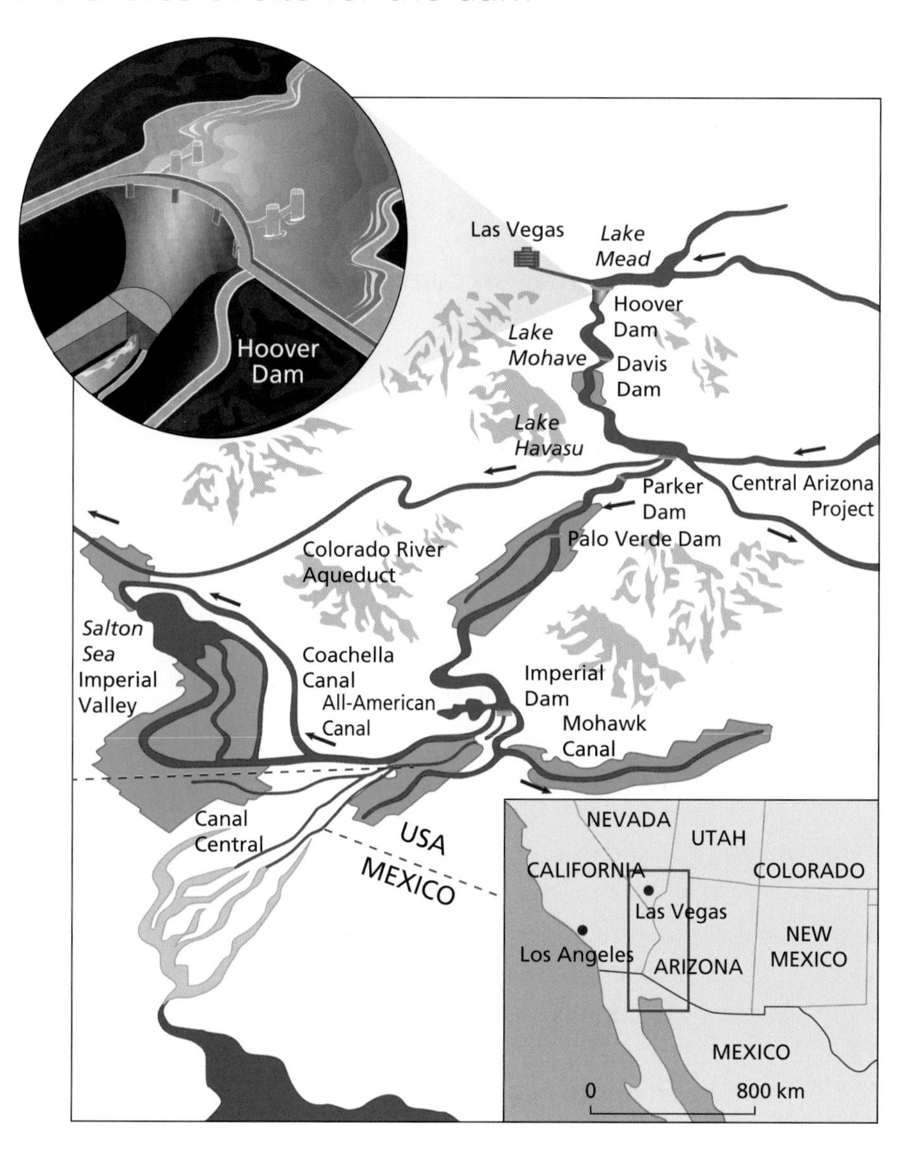


**Figure 3.7** Dams on the Colorado River, USA

**Physical factors:**

- The dams should have solid foundations to support their great weight.
- The Hoover Dam was built in a narrow valley to reduce the length of the dam and so reduce the costs.
- There was a large deep valley behind the dam suitable for the formation of a reservoir, which became Lake Mead.
- The rock beneath the reservoir and the dam needs to be impermeable to reduce water loss through seepage.
- The surface area of the reservoirs should be as small as possible to reduce evaporation.
- The catchment area needs to produce enough water to fill the reservoirs.

**Human factors:**

- The dams should be built in areas that are accessible for workers and materials.
- The desert cities of Las Vegas and Phoenix are expanding and close enough to provide a market for the electricity supplied from the dams.
- The amount of people displaced and the amount of farmland lost should be taken into account when locating dams and reservoirs. The cost of payments for relocating people and farmers needs to be considered.
- Flooding historical sites, for example the Rainbow Bridge, should be avoided if possible, as should sacred burial ground.
- The location of dams and reservoirs should take into account disrupting present transport routes.

## Impacts and consequences of the scheme

**Economic advantages:**

- The dams produce cheap hydroelectric power, which attracts industry into the area, providing jobs and increasing the standard of living.
- The water and power produced has encouraged the expansion of the cities of Las Vegas and Phoenix.
- The facilities in these cities attract many tourists, bringing money into the economy.
- The availability of irrigation water for farming means agricultural produce increases. This extra produce for sale creates more income for the economy.

**Economic disadvantages:**

- The schemes cost a huge amount of money to build and maintain.
- The more irrigation water used by farmers along the river the more saline the water becomes. When the water eventually reached Mexico it was unusable. To solve this problem a huge desalination plant had to be built at Yuma, which was expensive to build and maintain. The plant cost over $300 million to build and costs around $20 million a year to run.
- Water is very cheap for farmers to buy so much is wasted. The cost of delivering water to the farmers is around $350 per acre-foot but only costs the farmer $3.50 per acre-foot so they do not try to conserve it, which is a huge loss to the economy.
- Large amounts of compensation need to be paid out to people and farmers to be relocated as the reservoirs flood their land.

**Social advantages:**

- There is now an improved water supply for drinking as the dams ensure a constant supply.
- Irrigation water allows farmers, for example in California, to grow crops all year round. This increases food supply.
- The availability of water can sustain increasing populations, especially in the desert cities of Phoenix and Las Vegas. The Colorado River provides water for over 40 million people.
- There is a greater availability of electricity.
- The local populations around Lake Mead now have access to recreational activities like water sports, available because of the creation of the Lake Mead reservoir for the Hoover Dam.
- The Hoover Dam is a tourist attraction, bringing not only money into the area but also the facilities built for the tourists improve the social life of the local populations too.

**Social disadvantages:**

- Many people were forced to leave their homes to allow the flooding of the valleys to create the reservoirs. The town of St Thomas was drowned beneath the waters that created Lake Mead.
- The people had to be resettled in other areas, which were often not as productive as the land they left.

**Environmental advantages:**

- There is a constant supply of water for domestic use, which benefits health.
- The creation of reservoirs encourages wildlife into the area. More than 250 species of birds have been counted in the Lake Mead area.
- The reservoirs and dams are seen by some people as improving the scenery and environment.

**Environmental disadvantages:**

- The original wildlife in the area has been forced to move as its habitat has been destroyed by the reservoirs. There are no longer wild beavers in Tucson.
- The natural wild-scape has been destroyed by the dams and associated reservoirs.
- The level of Lake Powell is so high that the Rainbow Bridge, said to be one of the geological wonders of the world, is slowly being dissolved by the water.

**Political consequences:**

- The Colorado River runs through seven states of the USA and Mexico. All parties have to agree to the allocation of the Colorado water.
- Different states have different needs for domestic use, agriculture and industry.
- Each state can pass laws regarding water but these can conflict with each other so it needs the federal government to take control.
- Each state must agree to contribute to the cost and maintenance of the dams and reservoirs.
- The legal system in the USA worked in favour of the richest state (California) so it had more power over the water than the other states, which led to conflicts.

- Mexico was unhappy because, being the last area to receive the water, the quality of the water was poor and agreements were needed to sort out who paid for the cleaning of the water, for example the desalination plant at Yuma. Also, by the time the water reaches Mexico there may not be enough left to meet its allocation.

## Example

Describe the physical factors that need to be considered when selecting the site for a dam and its associated reservoir.

Explain the importance of these factors. **8 marks**

### Sample answer

*The site should have solid foundations (✓) to ensure it can support the great weight of the dam. (✓) The site should have impermeable rock (✓) as this means less water will be lost through seepage. (✓) It should have a narrow deep valley (✓) as this reduces the size of the dam so fewer building materials are needed, reducing the cost. (✓) The dam should be sited in an area away from earthquake zones (✓) to reduce the risk of the dam failing and collapsing. (✓) There needs to be an area of high rainfall (✓) to ensure there is a constant supply of water to fill the reservoir and stop the need to transfer water from other nearby river basins. (✓) There should be settlements close by to provide for workers. There should be settlements that could use the electricity provided by the dam.*

### Comments and marks

This is a very good answer, so much so that there are more ticks than the maximum mark allocation allows. This has been done to illustrate which points in the sample answer are mark-worthy. Marks are awarded for both description and explanation. The last two sentences gain no marks as they are human factors and do not refer to the site of the dam.

This answer gains **8 marks out of 8**.

## Key words and associated terms

**Drainage:** The term used to describe all surface water in a river system; not to be confused with underground pipes used to drain water from bogs or marshland.
**Infiltration:** The process by which water from precipitation seeps into the soil and subsoil.
**River basin:** The water catchment area of a river, including the main river and its tributaries.
**Salinity:** The salt content in surface water such as rivers, streams and lakes.
**Seasonal fluctuation:** When the normal climatic pattern is interrupted, for example when there is a change to rainfall distribution that could result in drought.
**Silting:** The amount of sand and other material carried in solution in rivers, which, when deposited, can reduce river flow.
**Tributary:** A smaller river or stream that runs into a larger river.
**Water control project:** The name given to efforts to manage various aspects of a river basin such as river flow, reservoirs, silting, navigation and river discharge through, for example, the use of dams, canals, aqueducts and diversion schemes.

Chapter 3.2

# Development and health

## Development indicators

### Description of countries as 'developed' or 'developing'

The level of development of a country can be determined by referring to a set of **economic and social indicators**.

Usually it is very difficult to define the level of development of a country based on one single indicator. It would be more accurate to refer to a combination of indicators.

*Key point* 

You should be able to explain differences in levels of development and explain the limits of some indicators, such as GNP in accurately reflecting different levels of the standard of living within any one country.

#### Economic indicators

- The Gross National Product (GNP) or Gross National Product statistics.
- Data relating to the relative percentage of the workforce employed in industry and agriculture.
- Data relating to average income per capita.
- The consumption of electricity per capita (in kilowatts per capita).
- The percentage of unemployment.
- Figures showing steel production in tonnes per capita.
- Trade patterns: import and export figures.
- Trade balances: surplus or deficits.

#### Social indicators

- Birth rates, death rates, infant mortality rates, life-expectancy rates.
- Population structure: the distribution of age and sex of the population.
- The average calorie intake per capita.
- The average number of people per doctor.
- Literacy rates (as an indication of the level of education).
- The percentage of the population with access to clean water.
- The percentage of the population who are homeless.
- The percentage of the population who attend primary school and who attend secondary school.

## Validity of development indicators

- The use of one individual factor to determine development could be misleading, especially if that factor is based on average figures such as income per capita or GNP, for example, those figures for Saudi Arabia may seem high and suggest a high level of development but income distribution is very unequal and varies from extremely high to very low.
- Similarly, GNP may indicate a high level but be based on a single commodity such as oil. This does not reveal the possible wide variations existing within a country, where some people may be very wealthy while others live at subsistence level.
- In some Middle Eastern oil-producing countries, GNP might appear to rank alongside those of highly developed countries, but this wealth is not evenly spread throughout the population.
- Caution must also be used when looking at some indicators relating to social and economic development levels in, for example, Brazil or India.
- Therefore using a combination of indicators, rather than individual indicators such as GNP, to produce a quality of life index is the best method of assessing levels of development in any given country.
- Taken together to produce combined indices such as the **Physical Quality of Life Index (PQLI)** or the Human Development Index (HDI), this gives a much more accurate picture of the stage of development of any given country.
- The Physical Quality of Life Index (PQLI) uses more than one **development indicator** and therefore gives a more balanced view of the level of development in a country.
- Indicators such as average life expectancy, adult literacy and infant mortality rates are used to calculate the PQLI.
- PQLI is measured on a scale of 0 to 100. If a country's PQLI is below 77, the country is said to be poorly developed.
- Average life expectancy is how long a person is expected to live. It can reflect the standard of living in a country. In the UK, the average life expectancy is 78, and in Sierra Leone it is less than 40.
- Literacy levels reflect the ability to read and write. In developed countries such as the UK, literacy levels are at 99 per cent, whereas in Sierra Leone it is only 35 per cent.
- High literacy rates indicate that the country has money to spend on schools, training teachers and buying materials such as books and equipment.
- In the UK it is compulsory for children to attend school between the ages of 4½ and 16. In poorer countries a smaller percentage of children attend school beyond primary level because they are needed to work on the land or to look after elderly family members.

*Key point* 

You should be able to identify different levels of development from given resources, such as a table or map, and be able to suggest suitable socio-economic indicators that could be used to produce maps that show, for example, developed and developing countries.

- Infant mortality rates are the number of children who die before their first birthday per 1000 of the population. It can reflect the level of health care available in a country. Rich countries can afford to build and resource hospitals and train doctors and nurses, more than poorer countries can.
- In the UK, infant mortality rates are low – about 5 per 1000 – compared to Sierra Leone, where figures are much higher at 155 per 1000.
- Taking all of these measures together gives a more accurate insight into levels of development, and makes it easier to compare countries. Single indicators can be too generalised and do not reveal variations within a country.

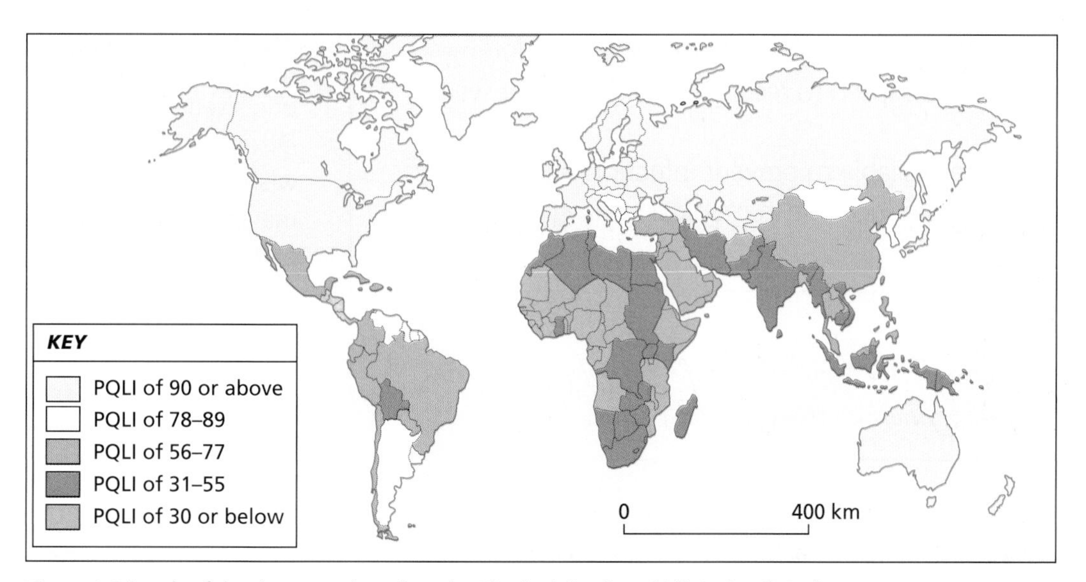


**Figure 3.8** Levels of development based on the Physical Quality of Life Index (PQLI)

# Differences in levels of development

- Countries such as Saudi Arabia, the United Arab Emirates and Brunei have prospered due to oil and gas reserves.
- Singapore, South Korea and Taiwan have encouraged the development of industry and commerce due to their entrepreneurial skills and have prospered.
- Countries such as Ethiopia or Chad lack natural resources and experience recurring drought that leads to famine.
- Some countries, for example Bangladesh, suffer natural disasters such as floods and cyclones.
- Political instability, problems of rapid population growth and civil disorder also affect economic growth in many developing countries.

What is the Physical Quality of Life Index (PQLI)?

# Water-related diseases

## Malaria

- Malaria is spread by vectors that carry the disease; namely, the female anopheles mosquito.
- They pick up the disease through taking blood meals from infected persons and pass it on in the next blood meal through their saliva.
- These mosquitoes breed in stagnant water, for example marshlands under certain climatic conditions; generally hot, wet climates with a minimum temperature of 16° C.
- The disease can spread very rapidly throughout an area unless certain measures are taken to limit and control this spread.
- Mosquitoes have become resistant to many insecticides, including DDT, and malaria itself has adapted to resist certain drugs that were formely used to cure it.
- As yet there is no vaccine available to prevent infection, although great efforts are being made to produce one in medical research facilities.

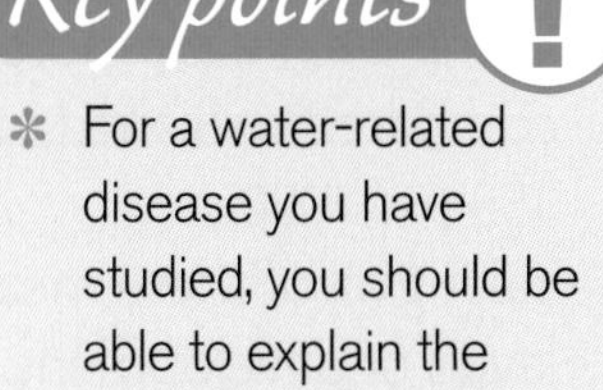

**Key points**

- For a water-related disease you have studied, you should be able to explain the causes and the impact, and the management strategies employed to control the disease, along with their effectiveness.
- Diseases you may wish to study include malaria, cholera and bilharzia/schistosomiasis.
- Your teacher will decide which disease(s) you will study for the examination.

**Hints & tips**

*Wherever possible in your answer, try to provide specific examples of countries by name and specific projects of health care that you have studied.*

### Impact of the disease

Malaria remains an important debilitating and killer disease in many parts of the developing world.

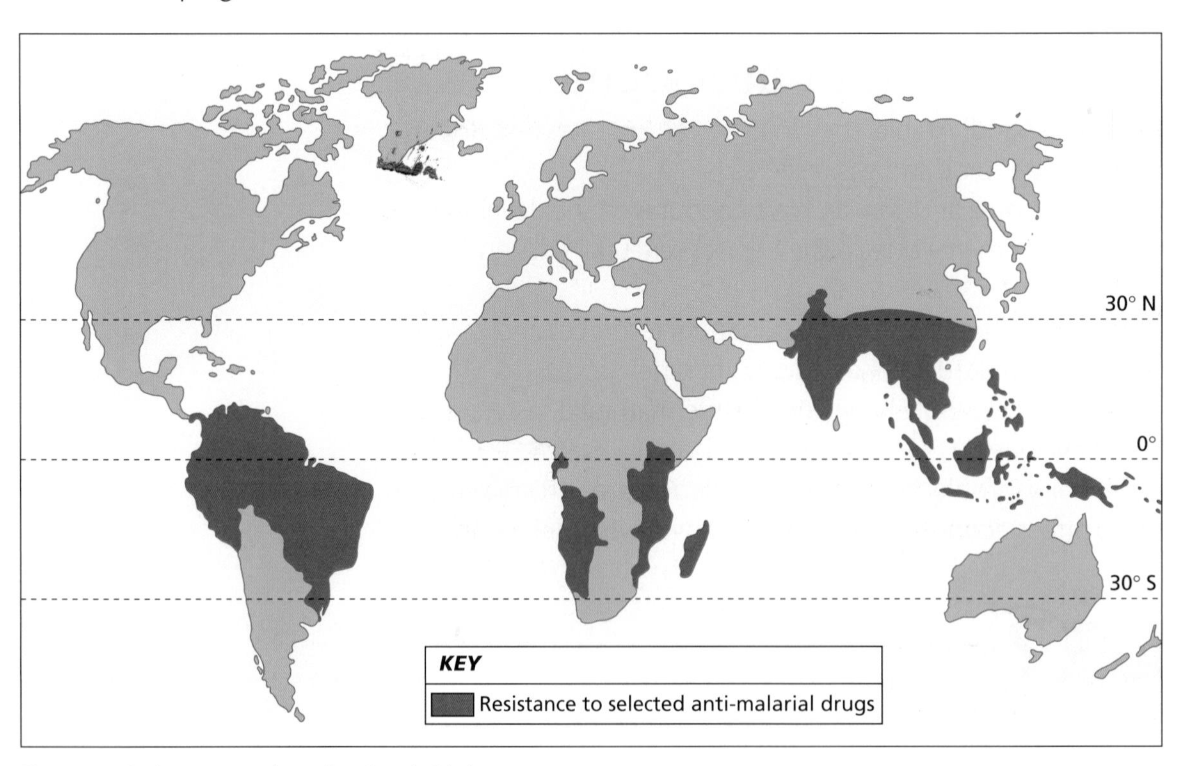


**Figure 3.9** Resistance to selected anti-malarial drugs

## Management strategies used to control the disease

- Using village health centres and issuing information/education through **primary health-care** schemes.
- Draining areas of stagnant water, for example swamps, and using water-management schemes to destroy the breeding grounds of mosquitoes near rivers.
- Using insecticides such as malathion.
- Covering the skin with clothing and using insect repellents.
- Using nets to protect people from mosquito bites while sleeping. Sleeping under nets treated with insecticide is an effective form of prevention because mosquitoes are most active between dusk and dawn.
- Using drugs to control the disease, for example quinine or derivatives of this drug such as chloroquinine.
- Releasing water from dams to drown larvae.
- Applying mustard seeds to water areas. The seeds become wet and sticky and drag the larvae below the water surface, which drowns them.
- Spraying egg white onto stagnant surfaces to suffocate mosquito larvae by clogging up their breathing tubes. This strategy is still at the experimental stage.
- Introducing small fish into paddy fields to eat larvae.
- Planting eucalyptus trees to absorb moisture.
- Infecting coconuts with BTI to eradicate mosquitoes as the larvae eat the bacteria, which destroy their stomach lining.
- Educating local populations to help reduce the number of cases, for example advising on the use of nets, insect repellents and insecticides on the nets, and the removal of anything containing stagnant water, such as water drums. These programmes are undertaken through primary health-care schemes.
- People can be issued with sprays and creams such as Autan, which can prevent bites.
- Anti-malarial drugs can be issued, but these can be expensive and must be taken regularly to be effective.

## Effectiveness of these strategies

- These measures have met with varying degrees of success.
- Much depends on local populations applying themselves to suggested precautions and taking medication regularly.
- Draining areas of stagnant water is time consuming and would have to be done regularly. The female anopheles mosquito only requires a small amount of still water to lay eggs in so this has not been very successful.
- DDT is harmful to the environment and is now banned in many areas. It has been replaced by malathion. Malathion is an oil-based insecticide that is more effective but more expensive than DDT, so many poor developing countries cannot afford it, and people do not want it sprayed in their homes as it stains the walls and does not smell very nice.
- If people cover their skin at times when the mosquitoes are most active they can stop themselves being bitten, reducing their chances of catching malaria.

- Insect repellents protect the skin as many contain deet, which the mosquitoes do not like.
- Mosquito nets are cheap and easy to use so they have been more successful than covering skin with clothing and using insect repellents.
- The drugs used to treat sufferers have had varying levels of success.
  - Chloroquinine is the cheapest of the three drugs available but mosquitoes have developed a resistance to it so it is no longer successful.
  - Lariam is very powerful and will help to protect people but it is not popular as it has harmful side effects.
  - Malarone is newer and has fewer side effects but is very expensive to produce so is not used in many poor countries, even though it is about 98 per cent effective.
  - Quinghaosu in pill form is seen by doctors and drug companies as the long-awaited breakthrough, but there is still not enough evidence to suggest that it will be really successful. The quinghaosu plant would have to be grown in large quantities and in areas outside China so that pills could be manufactured on a large scale.
- Applying mustard seeds and spraying egg white are deemed impractical and a waste of food sources.
- Adding fish to paddy fields has been quite successful as they are cheap and easy to breed. They can also be eaten, providing the locals with a valuable source of protein.
- BTI-infected coconuts are a successful way of eradicating mosquitoes. This method is cheap and environmentally friendly and a pond can be controlled for about 45 days using only a few coconuts. However, this can only be done in areas where there is a plentiful supply of coconuts and some people would say that it is also a waste of a valuable food source.
- Many people see vaccines as the way forward as children could be vaccinated at an early age and this could prevent them catching malaria. This could reduce infant mortality rates and would be successful in controlling the disease, provided people got themselves and their children vaccinated.
  - No effective vaccines have been produced, although several test studies in China, for example, are achieving progress.

# Cholera

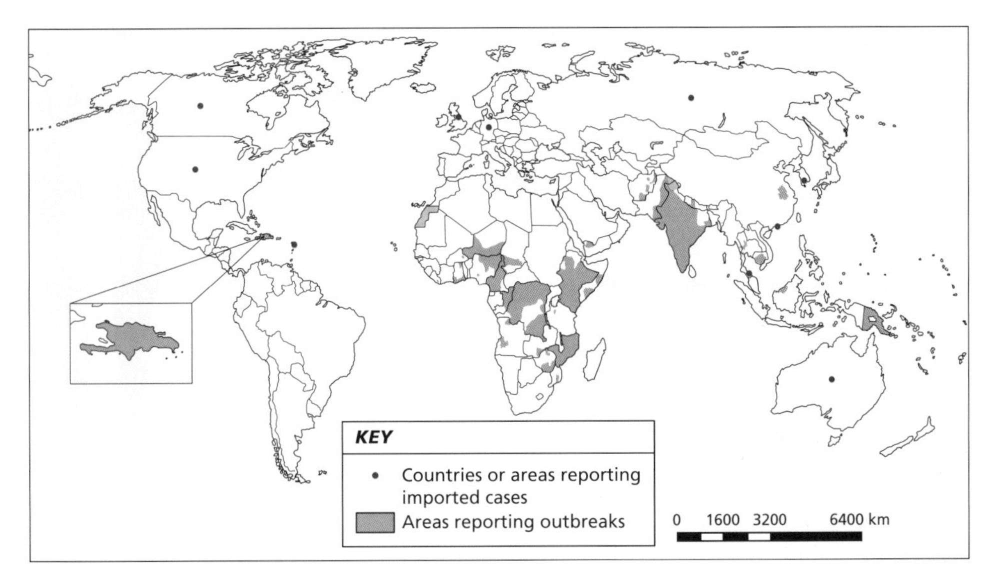


**Figure 3.10** Cholera outbreaks 2010–11

## Causes of the disease

- Cholera is caused primarily through transmission of bacteria, known as vibrio cholerae, from faecal contamination of food and water due to poor levels of sanitation.
- The disease can be spread by a lack of treatment of raw sewage, drinking-water supplies, bodies of water that can serve as reservoirs and by infected seafood that is shipped long distances.
- This bacteria infects the small intestine, which results in an acute diarrhoeal disease that can kill within hours if left untreated.

## Effects of disease

- The symptoms of cholera are profuse diarrhoea and vomiting of clear fluids. The disease has been nicknamed the 'blue death' because a sufferer's skin may turn bluish grey due to extreme loss of fluids.
- Severe diarrhoea can result in life-threatening dehydration and electrolyte imbalances. This leads to a drop in blood pressure, muscle cramping, unconsciousness and even coma. Children are more susceptible, especially two- to four-year-olds.
- Cholera affects an estimated 3 to 5 million people worldwide and causes between 58,000 and 130,000 deaths annually. The disease remains both epidemic and endemic in many areas of the world.

## Management strategies used to control the disease

- Treatment involves continued eating, which speeds the recovery of normal intestinal function. In most cases cholera can be successfully treated with oral rehydration therapy, which involves large volumes of fluids consisting of cooled boiled water, salt and sugar.
- Ten per cent of a person's body weight in fluid may need to be given in the first two to four hours.
- Antibiotics, such as doxycycline, can shorten the course of the disease and reduce the severity of the symptoms. So long as sufficient rehydration is maintained, people can recover without taking antibiotics.
- The disease can be successfully prevented if proper advanced water treatment and sanitation practices are followed. Sewage should be treated with antibacterial agents such as chlorine and ultraviolet light.
- All drinking water should be purified by boiling, chlorination and ozone water treatment. Governments need to fund these measures and have an effective surveillance system in place.
- All materials that come in contact with cholera patients should be sanitised by washing in hot water with chlorine bleach if possible. Personal hygiene is also important, especially washing hands when handling food.
- Warnings about possible cholera contamination should be posted around contaminated water sources.
- Vaccines such as Dukoral can be given to protect people from the disease.

# Primary health care in the developing world

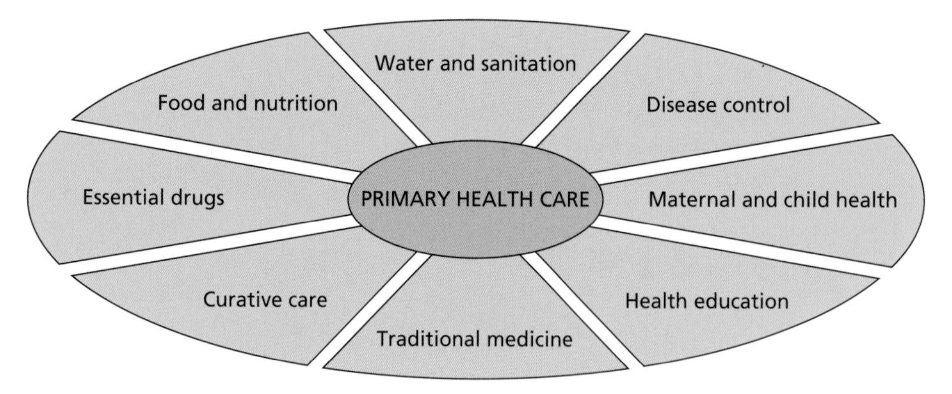


**Figure 3.11** Summary of primary health care

- Primary health-care (PHC) programmes are trying to provide villages with local dispensaries, which provide access to essential modern drugs, traditional remedies and family planning. Often these are supported by national policies, for example in India.
- Organisations are looking into how they can improve the sanitation facilities and the amount of clean drinking water available to people in rural areas, as improvements in these areas would lower cases of cholera and malaria.
- There is also a development of local health education programmes, especially in urban areas.
- There is a more widespread use of cheaper pharmaceuticals, rather than more expensive drugs, which can be just as effective.
- Generally these measures have been regarded as very effective, especially in reducing costs of medical care.
- Aid programmes that focus on primary health care are often more effective than any that provide large-scale medical programmes, such as the building of a large hospital that may treat only a small minority of cases.
- Successful primary health-care programmes improve the health of the local communities and have wider social benefits: if people are healthier they are more able to work and provide for their families.

## Barefoot doctors

- Primary health-care (PHC) strategies have been introduced by many developing countries in an effort to improve the health of the population. In rural areas of China the 'barefoot doctors' programme has been introduced.
- Barefoot doctors are local people who are given basic training so that they can attend to their communities' health needs.
- They treat simple ailments and minor injuries.
- Although the barefoot doctors cannot carry out major procedures, they can provide people with important information and services.
- They can provide families with advice on birth control, vaccinations and basic hygiene.
- As well as giving out advice, the medics can visit the sick who are too ill to travel to the nearest hospital.
- They can also run clinics in the larger villages and take a mobile health van to more remote villages.
- Routine medical situations are dealt with at local clinics.
- There are efforts to increase the development of a network of local clinics.

### Advantages

- Medical aides are available within easy reach of the community and this saves them travelling to the nearest hospital.
- It takes pressure off the large hospitals, allowing them to deal with more serious illnesses.
- Costs are reduced by referring only the most serious and complex cases to hospitals.

- The barefoot doctors are effective as they provide hope of health care for people in more remote areas where budgets and manpower can be limited.
- Training costs are low. In a similar scheme in India it costs about $100 to train a health worker for a year.
- Barefoot doctors are well known and trusted by villagers to provide them with vital health care. If the cases are too complex they are referred to the nearest hospital, where highly trained doctors and nurses can provide a wider range of facilities and medical care. These hospitals are very effective as they provide a good standard of medical care and money is invested in them by local governments.
- Barefoot doctors can provide advice on measures to help prevent the spread of disease and to reduce birth and infant mortality rates.
- The use of oral rehydration treatment to tackle diarrhoea is cheap and easily administered by barefoot doctors.

### Disadvantages

- However, despite the success of the barefoot doctors, there are not enough of them to care for everybody.
- Some lack the medical skills required and this leads to inappropriate prescribing of drugs and incomplete surgery.
- The system often breaks down if funding is no longer available for further training.

## Other strategies

- Other strategies include 'village reach programmes', which help to reduce the number of cases of malaria and diarrhoea in children. For example, in Malawi health workers trained through these programmes offer advice on preventing malaria and on general health and hygiene, and they have tried to improve the supply of clean water to villages. Wells and water pumps have been built and improved sewerage and sanitation systems have been installed.
- Health workers give advice to women on family planning and birth control.
- Health workers' advice to women can also help babies to survive, thereby reducing infant mortality rates.
- As in many African countries, AIDS is a major problem in Malawi and causes the deaths of people aged between 20 and 29. In some districts, 25 per cent of the adult population is HIV positive. Clinics operated by groups such Médicins Sans Frontières offer advice on safe sex and issue contraceptives and other birth-control devices.
- During 2014 and into 2015, several countries in West Africa, including Sierra Leone, experienced an outbreak of Ebola, a deadly disease. Up to 8000 people have died from this infection.
- Governments have employed local people to collect the bodies of victims, since the virus can exist after death, and take them to special burial areas. These workers have been provided with special protective clothing.
- Infected corpses and houses are sprayed with disinfectant chemicals.

- In addition to these efforts, many volunteers with medical training have travelled to this area from all over the world to assist local medical staff in treating and containing this epidemic. Despite precautions, some have contracted the disease and some have unfortunately died from it.
- Efforts are being made to find a vaccine for this disease, but none has as yet been found to be completely effective.

**Example** 

For a water-related disease you have studied, explain the human and environmental factors that can contribute to the spread of the disease. **10 marks**

## Sample answer

*The water-related disease I have chosen is malaria. The climate must have moderate rainfall and temperatures of 15 to 40 degrees C for the Anopheles mosquito to survive. (✓) Areas of stagnant water are needed for breeding grounds (✓) and vegetation for digestion of blood meals. (✓) There must be a human blood source nearby. (✓) If people are outside farming, particularly at dusk when mosquitoes are most active, they are at risk of being bitten. (✓) Pregnant women and young children are most at risk. If there is migration, so people are bringing the plasmodium parasite into the area, this increases the risk (✓) as does the lack of health knowledge and education. (✓)*

## Comment and marks

This is a good answer that shows understanding of the physical and human reasons for the spread of malaria. It is worth a total of **7 marks out of 10** available.

**Example**

For any water-related disease you have studied:

(i) Describe the measures that have been taken to manage the disease, and

(ii) Comment on the effectiveness of the measures. **20 marks**

## Sample answer

*The water-related disease I have chosen is malaria. Drugs such as chloroquine and artemisinin are used to combat the blood parasite. (✓) This is reasonably effective but the parasites can become resistant. (✓)*

*Insecticides such as DDT can be used to kill adult mosquitoes. (✓) This eradicated malaria in some areas but the mosquitoes have developed resistance and in some areas for example, Sri Lanka, the disease is returning. (✓)*

*Methods such as spraying ponds with egg white or adding mustard seeds to kill the larvae are not particularly effective and are wasteful. (✓) Adding fish, e.g. Nile Tilapia (✓) to ponds to eat larvae can reduce populations by 94% and add protein to people's diet. (✓) Coconuts impregnated with Bti bacteria (✓) are put into ponds. This destroys the gut of the larvae and is reasonably cheap, and two or three coconuts can control a pond for up to 45 days. (✓)*

*Flushing or draining ponds to kill the larvae is effective but it loses drinking water. (✓)*

*Education about the use of insecticide treated bed nets is a good idea but many people still use them incorrectly. (✓) People are also taught about sprays containing DDT to repel mosquitoes. This can easily be obtained through Primary Health Care.*

*Possible future methods include genetically modified male mosquitoes which are sterile, (✓) the use of a protein called Cectin which kills the parasite in the mosquito. (✓)*

## Comment and marks

This is a good answer that identifies correct methods of controlling the disease. The answer correctly identifies drugs to control the disease and also identifies insecticides that are used. It also notes the limitations of some insecticides. The latter part of the answer explains the use and relative effectiveness of a variety of methods used to control the spread of the disease and to provide protection for local communities. This answer has sufficient points to merit **13 marks out of 20** available.

## Key words and associated terms

**Development indicators:** Factors relating to the social and economic details of a country that can be used to demonstrate the level of its development.

**Physical Quality of Life Index (PQLI):** An index based on a combination of social and economic factors, which gives a more accurate picture than other indicators of the level of development of any given country.

**Primary health care:** A system designed to provide basic health and medical care to people in developing countries that is cost effective and readily available to people suffering from relatively minor health complaints. Rather than using highly trained medical staff or expensive hospitals, it relies on people who have basic medical skills and is therefore available to a larger number of the population.

# Physical and human causes of climate change

This section deals with physical and human causes of climate change and reasons for variations in global temperatures during the last 100–150 years.

## Physical causes of climate change

- Changes in the amount of solar energy given out by the Sun throughout time.
- Activities on the Earth including volcanic eruptions and variations in the amount of atmospheric gases present.
- Changes in the movement of the Earth in orbit. Slight shifts in the Earth's angle of tilt and the orbit pattern around the Sun have contributed to significant changes in the temperature pattern.
- Gases given off from rotting vegetation in tundra areas have affected global temperatures.
- Sunspots also increase temperatures.

## Suggested human causes

- The wide-scale burning of fossil fuels – due to road transport, power stations, heating systems, cement production, for example – and forested areas throughout the world has released various chemicals into the air, including sulphur dioxide, carbon monoxide and carbon dioxide.
- The increasing release of pollutants from traffic, rubbish dumps and other similar sources: **carbon dioxide** (**$CO_2$**), **fluorocarbons** (**CFCs**) and **nitrous oxide** (**$NO_2$**) are largely responsible for the greenhouse effect.
- The coolants used in fridges and air-conditioning systems create CFCs, which are safe in a closed system but can be released if appliances are not disposed of correctly.
- Nitrous oxides are released from vehicle exhausts and power stations.
- Trees, which give out valuable oxygen to the air, have been cut down and removed in great quantities, particularly in the rainforests where more carbon dioxide is present in the atmosphere and less is being recycled in photosynthesis due to deforestation.
- Increasing industrialisation has released air pollution from chimneys and factories.
- Methane is released from rice paddies, which are increasing to feed rapidly growing populations in Asian countries; from cows, which are bred to meet increasing global demand for beef; and from permafrost melting in Arctic areas due to global warming.
- Sulphate aerosol particles and aircraft contrails also contribute greenhouse gases.

# The local and global effects of climate change

## The greenhouse effect

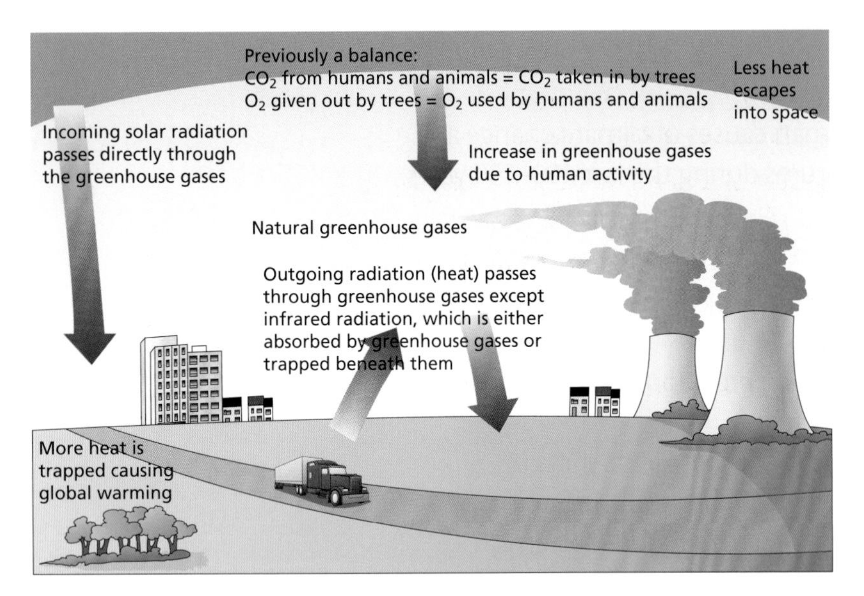


**Figure 3.12** The greenhouse effect

- Figure 3.12 summarises the factors involved in creating the greenhouse effect.
- Increase in gases such as methane from large herds of livestock, particularly cattle, contribute to the greenhouse effect.
- Release of CFCs (chlorofluorocarbons) from, for example, aerosols and refrigerants cause increases in global temperatures.
- The testing of atom bombs, which release radioactive material, may have affected the Earth's atmospheric conditions.

A number of people believe that many of the human reasons listed above have contributed to a process known as **global warming**. In effect, temperatures throughout the world are thought to have risen slightly in recent years.

## Effects of global climate change on the Earth

- Melting ice caps at the poles.
- A rise in sea levels caused by an expansion of the sea as it becomes warmer and also by the melting of glaciers and ice caps in Greenland and Antarctica; low-lying coastal areas, for example Bangladesh, are affected with the large-scale displacement of people, loss of land for farming and destruction of property.
- Changes in climates throughout the world: some places are experiencing milder winters and wetter summers.
- More extreme and variable weather, including floods, droughts, hurricanes and tornadoes becoming more frequent and intense.

- An increase in the process of desertification in arid and semi-arid areas.
- Globally, an increase in precipitation, particularly in the winter in northern countries such as Scotland, while some areas like the US Great Plains may experience drier conditions.
- Increased flooding due to increased rainfall.
- An increase in the extent of tropical diseases, for example yellow fever, as warmer areas expand, and up to 40 million more people in Africa exposed to the risk of contracting malaria.
- Longer growing seasons in many areas in northern Europe, for example, increasing food production and the range of crops being grown.
- Impacts on wildlife include the extinction of at least 10 per cent of land species and 80 per cent of coral reefs suffering bleaching.
- Changes to ocean current circulation, for example the thermohaline circulation in the Atlantic starts to lose its impact on north-west Europe, resulting in considerably colder winters.
- Changes in atmospheric patterns that link to monsoons such as El Niño and La Niña.
- An increase in cloud formation increases reflection/absorption in the atmosphere and therefore cooling.

## Disagreements over local and global climate change

Not all scientists agree on the concept of global climate change.

- It is generally assumed by the vast majority of scientists that global warming exists and that it has a dangerous impact on the Earth.
- However, many scientists have questioned the evidence for global warming, especially for human contributions to the causes.
- They have questioned the assertion that average global temperatures increased at a dangerously fast rate over the last few years. They have suggested that average global temperatures have only increased by 1–2°C, which they say is well within natural rates of climate change.
- Arguments in favour of global warming suggest that a rise of 2°C could lead to catastrophic effects, such as disruptions to ecosystems, an increase in storms, melting of ice in polar regions and a dangerous rise in sea levels.
- Arguments against this refute these changes, pointing out that the ice levels around Greenland and Antarctica attained a record area in 2007.
- They argue that warming due to carbon dioxide amounts to less than 1°C and that carbon dioxide is a beneficial fertiliser for plants and crops and aids efficient evapotranspiration.
- They quote meteorological experts who are agreed that there has been no increase in storms beyond natural variations in climate systems.
- Despite expenditure of $50 billion since 1990, many maintain there is no clear evidence that climate change is due to human causes.

- Many countries, with the exception of China and the USA during the second Bush administration, signed up to the Kyoto Protocol to limit harmful gases.
- It is suggested that this agreement will cost trillions of dollars and will adversely affect economies that can least afford to contribute to it, such as developing-world countries.
- The debate between those who support the arguments for measures to curtail global warming and those who argue against the existence of global warming will continue for many years.

# Management strategies employed to deal with climate change

## Efforts to manage global warming

- Countries pass laws to reduce the levels of emissions of harmful gases. The UK, for example, has introduced **green taxes** to reduce emissions from road traffic; from industries that emit gases such as carbon dioxide, sulphates and nitrates; and to reduce gases from rubbish tips and emissions from aircraft.
- Various countries have signed international agreements to reduce greenhouse gases, notably the Kyoto Protocol. Some of the countries, for example the USA, refused to sign this agreement during the second Bush administration.
- Legal measures have been taken in countries such as Brazil, where the rainforest is destroyed by burning trees to make way for cattle ranching.
- The USA has tried to persuade people to reduce their use of fossil fuels, notably oil, coal and gas, and replace them with less harmful alternative fuels.
- Advice on solutions to avoid global warming also includes reducing chemical fertilisers and insecticides by using compost or natural waste, recycling and using energy-efficient sources such as energy-saving light bulbs.
- Some governments have introduced, and monitor, better methods to dispose of industrial waste by removing harmful substances from it.

## Success of these efforts

- These measures have been successful up to a point.
- There are many with vested interests who have attempted to limit the use of these methods, especially oil companies throughout the world.
- Political lobbyists, working on behalf of these multinational companies, have persuaded governments to retain fossil fuels as the main source of energy.
- A significant number of scientists have rejected the arguments in favour of the harmful effects of climate change and instead have put forward counterarguments against these effects.

# Variations in global temperatures

You should be able to describe variations in global temperatures over 100 years as shown on a graph, for example. When describing the variations on the graph in Figure 3.13, you should refer to the following:

- Reference the general trend throughout the period, mentioning whether the temperature is generally rising or falling.
- Note any minor fluctuations within the graph and when they occur.
- Describe the highest and lowest points on the graph and the overall range in temperature change.
- State the periods when temperatures were below and above the stated norm, for example 1960.
- Note any times when the temperature remained fairly constant.
- Describe any periods when there were dramatic changes in the temperature.
- Refer to actual figures from the graph to illustrate your points.

**Key point**

You should be able to describe variations in global temperatures as shown on a graph, for example.

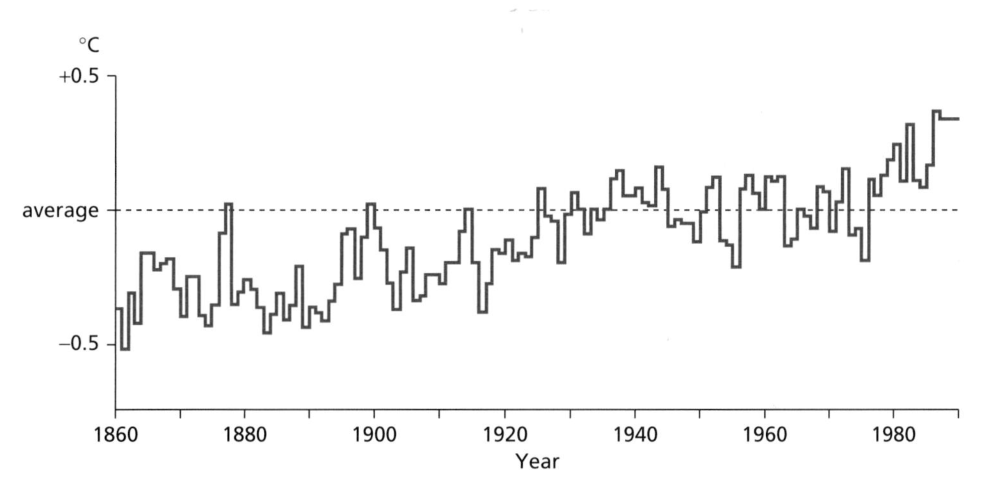


**Figure 3.13** Temperature variations from 1860 to 1990

# Physical reasons for recent variations in global temperatures

- Changes in the amount of solar energy given out by the Sun throughout time.
- Activities on the Earth including volcanic eruptions and variations in the amount of atmospheric gases present, which may have changed the amount of solar radiation that reached the surface of the Earth or have been absorbed in the atmosphere.
- Changes in the movement of the Earth in orbit. Slight shifts in the Earth's angle of tilt, the orbit pattern around the Sun and changes in the gravitational influences of the Sun and Moon on the Earth have contributed to significant changes in the temperature pattern.
- Gases given off from rotting vegetation in tundra areas have affected global temperatures.

**Key point** 

You should be able to suggest both physical and human reasons for variations in global temperatures for the last 100 to 150 years.

# Human reasons for recent variations in global temperatures

- The testing of atom bombs releasing radioactive material may have affected the Earth's atmospheric conditions.
- The wide-scale burning of fossil fuels and forested areas throughout the world has released various chemicals into the air, including sulphur dioxide, carbon monoxide and carbon dioxide.
- Trees, which give out valuable oxygen to the air, have been cut down and removed in great quantities. This has inevitably disrupted the balance of gases within the atmosphere and affected the processes involved in the Earth's insolation patterns.
- Increasing industrialisation releases air pollution from chimneys and factories.
- The increasing release of pollutants from traffic, rubbish dumps or other similar sources.
- Increase in gases such as methane from large herds of livestock, particularly cattle.

Those factors listed above have greatly contributed to the phenomenon known as the **greenhouse effect**.

- The gases responsible for the enhanced greenhouse effect include carbon dioxide, methane and fluorocarbons. The ozone layer lies about 10 to 50 kilometres above the Earth's surface and is very important since it acts as a shield against ultraviolet radiation from the Sun.
- Release of CFCs (chlorofluorocarbons) from, for example, aerosols and refrigerants, has reacted with gases in the ozone layer and has caused thinning or gaps known as the **ozone hole**.
- In addition to nitrogen and oxygen, there is a small but important number of gases that contribute to the process of heating and retaining heat within the atmosphere. These gases help to allow short-wave radiation back into space or absorb long-wave radiation leaving the Earth and re-radiate it back to the surface. Without these gases, temperatures at the Earth's surface could fall dramatically by perhaps 10–15° C. If the level of gases increases this could lead to a rise in temperature globally across the Earth.

The human consequences of this change in the atmosphere include increases in cancer and cataracts and reduced crop yields.

Draw a sketch to show the main elements of the greenhouse effect.

## Example

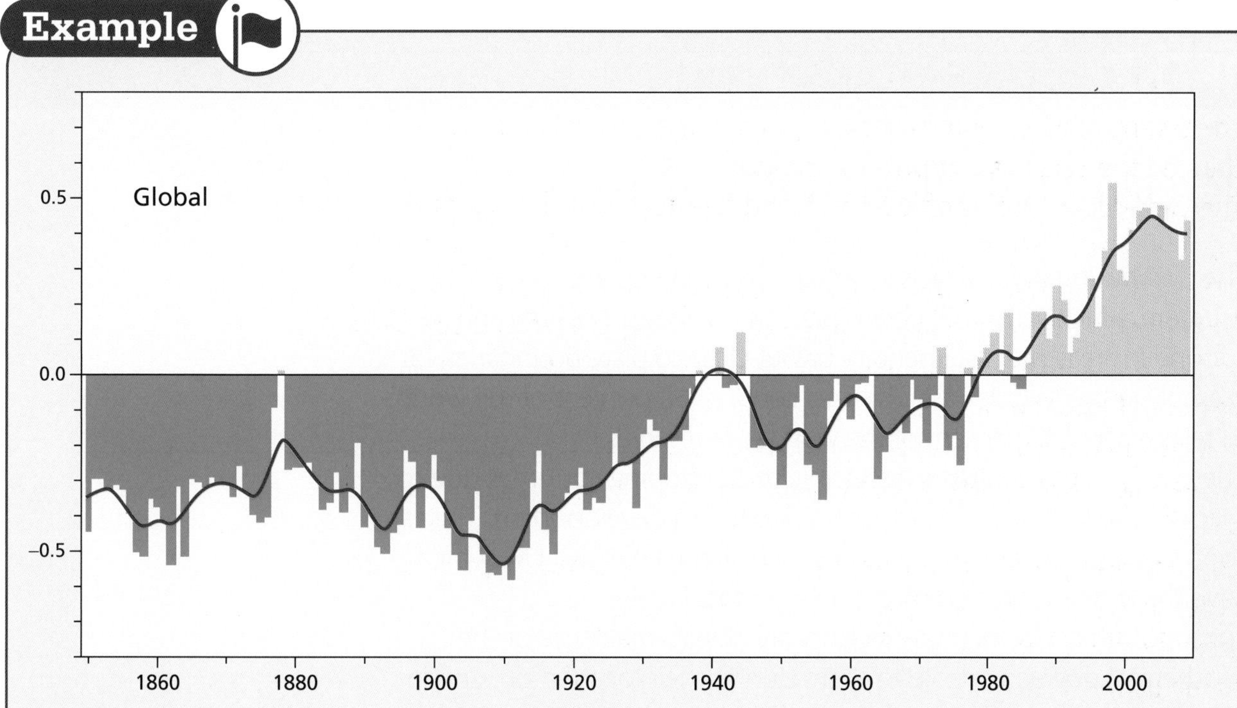


**Figure 3.14** Changes in global mean temperatures 1860 to 2000

Study Figure 3.14.

Discuss the physical factors and human factors that may have led to the changes in the global mean temperatures shown. **12 marks**

## Sample answer

*Physical factors which have led to global temperature change are the variation in solar energy for example sun spots, (✓) volcanic eruptions as they produce dust (✓) which absorbs solar energy therefore cooling the Earth. (✓)*

*Human factors are deforestation as the burning of trees causes $CO_2$ in the atmosphere. (✓) The burning of fossil fuels releases $CO_2$ into the atmosphere. (✓) CFCs found in refrigerators contribute to gases in the atmosphere. (✓) Methane produced by cattle and undrained marshlands contributes to harmful gases in the atmosphere. (✓) Rotting vegetation also releases harmful gases such as methane (✓) which contribute to an increase in the greenhouse effect and cause changes to global mean temperatures. (✓)*

## Comment and marks

This answer has sufficient points to achieve more than half marks; namely, **9 out of 12**. More marks could have been obtained had the answer had more detail on the physical factors and referred to industries and transport that burn fossil fuels.

## Key words and associated terms

For a glossary of associated terms, refer to the key words at the end of Chapter 1.1 Atmosphere (page 10).

# Chapter 3.4 Energy

## Global distribution of energy resources

- The distribution of fossil fuels is very uneven and depends on geological structure, which varies in different parts of the world. Most oil and gas is found in areas with sedimentary basins, close to plate boundaries, for example in the Middle East. This area contains 60 per cent of the world's oil reserves and 40 per cent of the world's gas reserves.
- These resources are often inaccessible to developing countries due to a lack of the technological expertise needed to develop them. This contrasts with developed countries such as the USA and UK, which have the money and expertise to extract fossil fuels.
- Some countries, for example Iceland, are able to make use of the geothermal power available to them due to their location on a plate boundary.
- Other countries benefit from their climate, such as Spain, which is able to use its sunny conditions to develop solar power. Areas with mountains and high rainfall have the resources to develop **hydro-electric schemes**.
- Recently, many developed countries with coastlines have harnessed the power of the sea and rivers to create power schemes.

## World energy demands

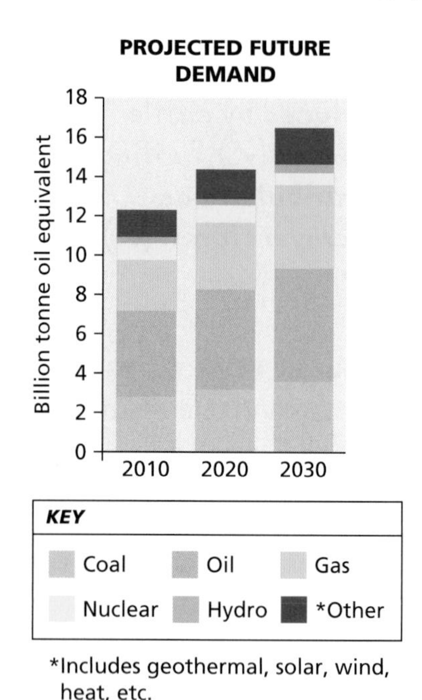


**Figure 3.15** World primary energy demand (projected)

Global demand for energy has increased over the last 150 years. By far the greatest demand for energy comes from developed countries in order to support their economies. Economic forecasters suggest that will change. It is thought that energy demands from developing countries will increase at a faster rate than developed countries. Demand is slowing down in developed countries as less manufacturing takes place and the emphasis shifts to service and quaternary industries.

Most of the global economic and population growth is happening in developing countries. Countries such as China need more energy to satisfy their economic growth; it is predicted that this will result in a rise in demand of 40 per cent. As their population continues to rise, so too will their energy demands, and as standards of living increase energy demand for electricity, electrical goods and transport will also increase.

Throughout the world there is an ever-increasing demand for both **non-renewable** and **renewable energy sources**.

This demand comes from:

## Residential use

- Electricity is needed for lighting and appliances such as televisions, washing machines and perhaps air conditioning.

## Industrial use

- A great deal of economic growth in developing countries is based on energy-hungry manufacturing industries. This accounts for some of the increased energy requirements in these countries.

## Transport

- Manufactured products that are sold to developed countries need to be transported around the world, using energy such as diesel and petrol.
- As people in developing countries become more prosperous, car-ownership rates will increase, resulting in more energy usage.
- In developed countries the population growth rates are more stable (or even declining), and so there is no great increase in demand for energy.
- New products and technologies are increasingly energy efficient, which keeps energy consumption in developed countries at a steadier rate.

*Key point* 

You should be able to discuss the varying demands for energy in developed and developing countries and the sources of the energy supplied to meet these demands.

# Non-renewable and renewable energy sources

## Non-renewable energy sources

- **Fossil fuels** such as oil (including shale oil and gas), coal, natural gas, biomass (from decaying plant or animal waste and other organic material), wood (from felling trees) and **nuclear** (from mined radioactive minerals such as uranium).

- These fuel sources, with the exception of wood and biomass, will eventually run out, and therefore relying on them is unsustainable.

**Advantage**

- The main advantage of these fuels is that they are relatively cheap to extract and to convert into energy.

**Disadvantages**

- When burned, fossil fuels give off atmospheric pollutants, including greenhouse gases.
- Nuclear plants are very expensive to operate and nuclear waste is highly toxic and needs to be stored for many thousands of years.
  - Storage costs are extremely expensive and any leakages can have a devastating effect on people and the environment.
  - The worst nuclear accident occurred in 1986 in Chernobyl in Ukraine.

Key point 

You should know about renewable and non-renewable energy sources and their respective advantages and disadvantages.

| Estimated length of time left for fossil fuels | |
|---|---|
| **Fuel** | **Time left** |
| Oil | 50 years |
| Natural gas | 70 years |
| Coal | 250 years |

# The effectiveness of renewable sources of energy

## Hydroelectric power

- Hydroelectric power is more effective where there is high rainfall, to ensure that reservoirs are always at capacity, and where there is suitable underlying geology (impermeable rock), to ensure water is not lost from the reservoirs through seepage.
- Hanging valleys often make ideal sites for effective hydroelectric power plants because they allow the vertical drop of water needed to power turbines.
- Hydroelectric pump storage schemes allow electricity to be instantly produced as required, and are currently being used to meet periods of peak demand throughout the UK and other European countries, the USA and countries throughout Africa and Asia.
- There are also hydroelectric power schemes that are based on river flow, such as the scheme on the River Rhine in Germany, and tidal schemes, such as the River Thames Tidal Scheme in London.
- These water-management schemes also provide protection from floods.

## Wind power

- Wind power is most effective where there are no barriers to the prevailing wind so allowing regular and reliable movement of air to turn the turbines.

**Advantage:**

- Once built, the energy produced is cheap.

**Disadvantages:**

- Concerns have been raised about how to bridge the energy gap when the wind turbines are not generating electricity on calm days, because it is difficult to store the electricity produced by them.
- Other drawbacks to this type of energy for developing countries include the initial high costs of construction.

## Wave power

- Wave power approaches are currently being developed, and are most effective in areas such as the Pentland Firth where the fetch is large.

## Solar heating

- Solar heating, using heat from the Sun through the use of solar panels in houses and industrial areas, is becoming much more common.

**Advantages:**

- Once installed, the energy produced is relatively cheap.
- Costs can be reduced by returning excess energy to national energy grids.

**Disadvantages:**

- Problems can arise in areas where sunshine is limited.
- The costs of manufacturing and installation are initially expensive.

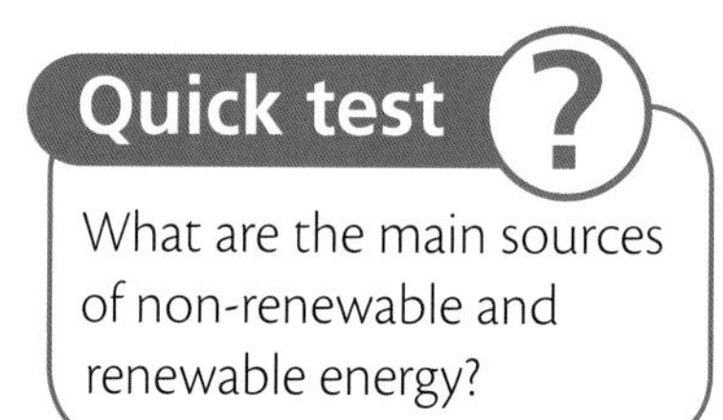

What are the main sources of non-renewable and renewable energy?

## Geothermal

- Geothermal sources produce heat in places where the magma layer is close to the surface of the Earth, such as in Iceland.

## Biomass energy and biofuels

- Biomass energy and biofuels can provide continuous energy as required by the burning of plant matter.

**Advantage:**

- As the carbon dioxide released equals what the plants recently took in, biomass energy does not add new greenhouse gases, and so is more environmentally friendly than burning fossil fuels.

**Disadvantage:**

- Drawbacks include concerns about air pollution increasing in the local area and using land that is needed for crop production in developing countries.

## Example

Explain the reasons for the increase in energy consumption in developing areas of the world. You should refer to examples in your answer. **10 marks**

### Sample answer

The population of countries like China and India is rising and increasing numbers of people need food and shelter so increase the demand for energy. (✓) In developing countries an increase in standards of living means an increase in the demand for energy, (✓) as many people now can afford televisions, washing machines, fridge freezers, air conditioning etc. (✓) Countries like China are developing manufacturing industries, which use up enormous amounts of energy especially fossil fuels. (✓) Developing countries increasingly trade bulky primary products with developed countries and many of these goods are transported around the world using up more fossil fuels. (✓) In countries like India, car ownership is becoming more common due to an increased standard of living, so more energy is needed to run these vehicles. (✓) Energy increase is less in developed countries because their populations are growing much more slowly. Developed countries are being encouraged to conserve energy to reduce the effects of global warming, so energy consumption is growing much slower.

### Comments and marks

This is a good answer. It mentions specific countries and gives reasons for the increase demand for energy. No marks are awarded for the last two sentences as these refer to developed countries.

This answer would be awarded **6 marks out of 10**.

## Key words and associated terms

**Energy source:** The type of fuel used to generate energy for industry, housing and transport.

**Hydroelectric schemes:** Schemes based on damming lakes to create electricity through the use of turbines powered by water.

**Non-renewable energy:** Energy that uses finite resources such as oil, gas and coal.

**Renewable energy:** Energy created from sources that can be renewed and that will not eventually run out, for example solar, tidal, wood and geothermal sources.